Years 9–10 Maths

FOR STUDENTS

Years 9–10 Maths FOR STUDENTS

by Ingrid Kemp
Mary Jane Sterling
Christopher Danielson
Mark Ryan
Mark Zegarelli

Made by the people who make the *DUMMIES* books!

A Wiley Brand

Years 9–10 Maths for Students®

Published by
Wiley Publishing Australia Pty Ltd
42 McDougall Street
Milton, Qld 4064
www.dummies.com

Copyright © 2016 Wiley Publishing Australia Pty Ltd

The moral rights of the authors have been asserted.

National Library of Australia
Cataloguing-in-Publication data:

Author:	Mary Jane Sterling
Contributors:	Ingrid Kemp, Christopher Danielson, Mark Ryan, Mark Zegarelli
Title:	Years 9–10 Maths for Students
ISBN:	9780730326779 (pbk.)
	9780730326793 (ebook)
Series:	For Dummies
Notes:	Includes index.
Subjects:	Mathematics — Textbooks.
	Mathematics — Study and teaching (Secondary)
	Mathematics — Problems, exercises etc.
Dewey Number:	510.76

Cover: Wiley Creative Services

Illustrations by Wiley, Composition Services Graphics

Typeset by diacriTech, Chennai, India

Printed in Australia by
Ligare Book Printer

10 9 8 7 6 5 4 3 2 1

Contents at a Glance

Introduction 1

Part I: Reviewing the Basics 5

Chapter 1: Assembling Your Tools 7
Chapter 2: Working with Whole Numbers 27
Chapter 3: Ups and Downs: Positive and Negative Numbers 49
Chapter 4: Parts of the Whole: Fractions, Decimals and Percentages 61
Chapter 5: Understanding Order of Operations 93

Part II: Algebra is Part of Everything 109

Chapter 6: Understanding the Basics of Algebra 111
Chapter 7: Working with the Variability of Variables 139
Chapter 8: Smaller is Better: Factoring Down 167
Chapter 9: Going for the Second Degree with Quadratics 191

Part III: Solving Algebraic Equations 211

Chapter 10: Establishing the Ground Rules and Solving Linear Equations 213
Chapter 11: Taking a Crack at Quadratic Equations 243

Part IV: Applying Algebra and Understanding Geometry 265

Chapter 12: Graphing Basics 267
Chapter 13: Graphing Lines, Gradients and Circles 279
Chapter 14: Getting Familiar with Functions 303
Chapter 15: Pythagoras, Trigonometry and Measurement 313
Chapter 16: Geometry Basics 337

Part V: The Part of Tens 365

Chapter 17: Ten Ways to Avoid Algebra Pitfalls 367

Index 373

Table of Contents

Introduction 1

About This Book 2
Foolish Assumptions 2
Icons Used in This Book 3
Where to Go From Here 3

Part I: Reviewing the Basics 5

Chapter 1: Assembling Your Tools 7

Starting with the Basics 8
Whole numbers: Adding, subtracting,
multiplying and dividing 9
Parts of the whole: Fractions, decimals and percentages 10
Moving On to Algebra 10
Speaking in Algebra 11
Taking aim at algebra operations 12
What About Geometry? 12
Playing with Maths 13
Experimenting with symbols 13
Building models 14
Arguing is heaps of fun 15
Connecting ideas 15
What Parents Can Do to Help 17
Focusing on asking questions 17
Helping your child with homework (without doing the
work yourself) 22
Becoming unstuck: What to do 24

Chapter 2: Working with Whole Numbers 27

Adding Things Up 27
In line: Adding larger numbers in columns................................ 28
Carry on: Dealing with two-digit answers................................. 28
Take It Away: Subtracting 31
Columns and stacks: Subtracting larger numbers...................... 32
Can you spare a ten? Borrowing to subtract.............................. 33

Multiplying..36
Signs of the times..37
Memorising the multiplication table 37
Double digits: Multiplying larger numbers 41
Doing Division Lickety-Split 43
Making short work of long division 44
Working through an example 45

Chapter 3: Ups and Downs: Positive and Negative Numbers 49

Showing Some Signs 49
Picking out positive numbers...50
Making the most of negative numbers..50
Comparing positives and negatives 51
Zeroing in on zero...52
Operating with Signed Numbers..52
Adding like to like: Same-signed numbers 52
Adding different signs 54
Subtracting signed numbers 54
Multiplying and dividing signed numbers 56
Working with Nothing: Zero and Signed Numbers 58

**Chapter 4: Parts of the Whole: Fractions, Decimals
and Percentages..61**

Multiplying and Dividing Fractions 62
Multiplying numerators and denominators straight across 62
Multiplying mixed numbers..64
Doing a flip to divide fractions 65
Dividing mixed numbers 66
All Together Now: Adding Fractions 67
Finding the sum of fractions with the same denominator............67
Adding fractions with different denominators.............................68
Taking It Away: Subtracting Fractions 75
Subtracting fractions with the same denominator.......................76
Subtracting fractions with different denominators.......................76
Performing the Main Four Operations with Decimals 80
Adding decimals..80
Subtracting decimals 82
Multiplying decimals 83
Dividing decimals..84
Checking your answers 88
Converting to and from Percentages, Decimals and Fractions..............89
Going from percentages to decimals...89
Changing decimals into percentages...90
Switching from percentages to fractions 90
Turning fractions into percentages 91

Chapter 5: Understanding Order of Operations.....................93

Ordering Operations 93
Applying order of operations to the main four expressions..........95
Using order of operations in expressions with
exponents and roots 98
Gathering Terms with Grouping Symbols 99
Understanding order of precedence in expressions
with parentheses...100
Putting it all together...103
Checking Your Answers 105
Making sense or cents or scents...105
Plugging in to get a charge of your answer 106

Part II: Algebra is Part of Everything.......................... 109

Chapter 6: Understanding the Basics of Algebra...................111

Looking at the Basics: Numbers 111
Really real numbers 112
Counting on natural numbers 112
Wholly whole numbers 113
Integrating integers...113
Being reasonable: Rational numbers.......................................113
Restraining irrational numbers 114
Picking out primes and composites 114
Deciphering the Symbols in Algebra Operations 114
Grouping 115
Defining relationships 116
Taking on algebraic tasks...116
Associating and Commuting with Expressions.................................117
Reordering operations: The commutative property..............117
Associating expressions: The associative property..............118
I Got the Power! Using Exponents 120
Understanding what exponents are 120
The first index law 121
The second index law..124
Getting Complicated with Exponents 126
The third index law: The power of zero...................................126
The fourth, fifth and sixth index laws: Powers of powers............126
The seventh index law: Working with negative exponents 128
The eighth index law 129
Comparing with Exponents 131
Taking notes on scientific notation...132
Exploring exponential expressions 133

Chapter 7: Working with the Variability of Variables **139**

Representing Numbers with Letters 140
    Attaching factors and coefficients 141
    Interpreting the operations 141
Doing the Maths 142
    Adding and subtracting variables 143
    Adding and subtracting with powers 144
Multiplying and Dividing Variables 145
    Multiplying variables 145
    Dividing variables 146
    Doing it all 147
Expanding Expressions 149
    Getting your equal share 149
    Distributing first 150
    Adding first 151
Distributing Signs 152
    Distributing positives 152
    Distributing negatives 153
    Reversing the roles in distributing 153
Mixing It Up with Numbers and Variables 154
    Negative exponents yielding fractional answers 156
    Working with fractional powers 157
Binomials and Trinomials: Distributing More Than One Term 159
    Distributing binomials 159
    Distributing trinomials 160
    Multiplying a polynomial by another polynomial 161
Making Special Distributions 162
    Recognising the perfectly squared binomial 162
    Spotting the sum and difference of the same two terms 163

Chapter 8: Smaller is Better: Factoring Down **167**

Beginning with the Basics 168
Composing Composite Numbers 169
Writing Prime Factorisations 170
    Dividing while standing on your head 170
    Getting to the root of primes with a tree 171
    Wrapping your head around the rules of divisibility 172
Getting Down to the Prime Factor 174
    Taking primes into account 174
    Pulling out factors and leaving the rest 177
Getting to First Base with Factoring 179
    Factoring out numbers 180
    Factoring out variables 182
    Unlocking combinations of numbers and variables 183
    Changing factoring into a division problem 185
Grouping Terms 186

Chapter 9: Going for the Second Degree with Quadratics **191**

The Standard Quadratic Expression 192
Reining in Big and Tiny Numbers 193
FOILing 194
    FOILing basics 194
    FOILed again, and again 196
    Applying FOIL to a special product 198
UnFOILing 199
    Unwrapping the FOILing package 199
    Coming to the end of the FOIL roll 203
Making Factoring Choices 204
    Combining unFOIL and the greatest common factor 204
    Grouping and unFOILing in the same package 206
Factoring the Difference of Two Perfect Squares 207
    Ending with binomials 208
    Knowing when to quit 209

Part III: Solving Algebraic Equations 211

Chapter 10: Establishing the Ground Rules and Solving Linear Equations **213**

Creating the Correct Setup for Solving Equations 214
Keeping Equations Balanced 214
    Balancing with binary operations 215
    Squaring both sides and suffering the consequences 217
    Taking a root of both sides 218
    Undoing an operation with its opposite 218
Solving with Reciprocals 219
Making a List and Checking It Twice 221
    Doing a reality check 221
    Thinking like a car mechanic when checking your work 223
Finding a Purpose 223
Solving Linear Equations: Playing by the Rules 224
Solving Equations with Two Terms 225
    Devising a method using division 225
    Making the most of multiplication 227
    Reciprocating the invitation 229
Extending the Number of Terms to Three 230
    Eliminating the extra constant term 230
    Vanquishing the extra variable term 231
Simplifying to Keep It Simple 233
    Distributing first 233
    Multiplying or dividing before distributing 235

Featuring Fractions 237
    Promoting practical proportions...237
    Transforming fractional equations into proportions....................239
Solving for Variables in Formulas..241

Chapter 11: Taking a Crack at Quadratic Equations............243

Squaring Up to Quadratics..244
Rooting Out Results from Quadratic Equations 246
Factoring for a Solution 249
    Zeroing in on the multiplication property of zero.......................249
    Assigning the greatest common factor and multiplication
       property of zero to solving quadratics..............................250
Solving Quadratics with Three Terms 252
Applying Quadratic Equation Solutions 257
Figuring Out the Quadratic Formula..259

Part IV: Applying Algebra and Understanding Geometry 265

Chapter 12: Graphing Basics 267

The Cartesian Plane 268
Grappling with Graphs..269
    Making a point..269
    Ordering pairs, or coordinating coordinates 270
Actually Graphing Points...272
Graphing Is Good...273
Graphing Formulas and Equations 274
    Lining up a linear equation..274
    Going around in circles with a circular graph.............................275
    Throwing an object into the air 276

Chapter 13: Graphing Lines, Gradients and Circles...............279

Graphing a Line...279
Graphing the Equation of a Line 281
Investigating Intercepts 284
Sighting the Gradient 285
    Formulating gradient..287
    Combining gradient and intercept..289
    Getting to the gradient-intercept form......................................290
    Graphing with gradient-intercept 290
Marking Parallel and Perpendicular Lines 292
Intersecting Lines and Simultaneous Equations.................................293
    Graphing for intersections...293
    Substituting to find intersections 294

Eliminating to find intersections 296
Applications of simultaneous equations 298
Working Out Distance and the Midpoint 299
The distance formula 299
The midpoint formula 300
Equations for Circles 300

Chapter 14: Getting Familiar with Functions 303

Curling Up with Parabolas 303
Trying out the basic parabola 304
Putting the vertex on an axis 305
Sliding and multiplying 305
Delving into Functions 308
Understanding the practical side of functions 309
Figuring out a function's function 310
Studying Function Families 310

Chapter 15: Pythagoras, Trigonometry and Measurement 313

Measuring Up 313
Finding out how long: Units of length 314
Putting the Pythagorean theorem to work 314
Working around the perimeter 316
Spreading Out: Area Formulas 320
Laying out rectangles and squares 321
Tuning in triangles 322
Going around in circles 324
Using area formulas for different shapes 324
Working with composite shapes 326
Pumping Up with Volume Formulas 327
Prying into prisms and boxes 327
Cycling cylinders 328
Scaling a pyramid 328
Pointing to cones 329
Rolling along with spheres 329
Triggering Trigonometric Ratios 330
Finding lengths 330
Finding angles 333
Understanding degrees and minutes 334

Chapter 16: Geometry Basics 337

Geometry Proofs 337
Am I Ever Going to Use This? 338
When you'll use your knowledge of shapes 338
When you'll use your knowledge of proofs 339
Getting Down with Definitions 339
A Few Points on Points 342

Lines, Segments and Rays 342
    Horizontal and vertical lines 343
    Doubling up with pairs of lines 343
Investigating the Plane Facts 344
Everybody's Got an Angle 345
    Five types of angles 345
    Angle pairs 346
Bisection and Trisection...347
    Segments 347
    Angles 348
Taking In a Triangle's Sides 349
    Scalene triangles 349
    Isosceles triangles..350
    Equilateral triangles 350
Proving Triangles are Congruent 350
    SSS: The side-side-side method..351
    SAS: side-angle-side 353
    ASA: The angle-side-angle tack 355
    AAS: angle-angle-side 356
    Last but not least: RHS 356
Similar Figures...357
    Defining similar polygons 357
    How similar figures line up 358
    Solving a similarity problem 360
Proving Triangles Similar..362
    Tackling an AA proof 362

Part V: The Part of Tens.. 365

Chapter 17: Ten Ways to Avoid Algebra Pitfalls 367
Keeping Track of the Middle Term 367
Distributing: One for You and One for Me 368
Breaking Up Fractions (Breaking Up Is Hard to Do).....................368
Renovating Radicals 369
Order of Operations 369
Fractional Exponents 369
Multiplying Bases Together 370
A Power to a Power 370
Reducing for a Better Fit...371
Negative Exponents..371

Index 373

Introduction

In this book, I offer a refresher on some basic maths operations, such as addition, subtraction, multiplication and division, before moving on to the more advanced topic of algebra. So let me introduce you to algebra. This introduction is somewhat like what would happen if I were to introduce you to my friend Donna. I'd say, 'This is Donna. Let me tell you something about her.' After giving a few well-chosen tidbits of information about Donna, I'd let you ask more questions or fill in more details. In this book, you find some well-chosen topics and information, and I try to fill in details as I go along.

As you read this introduction, you're probably in one of two situations:

- You've taken the plunge and bought the book.
- You're checking things out before committing to the purchase.

In either case, you'd probably like to have some good, concrete reasons why you should go to the trouble of reading and finding out about algebra.

One of the most commonly asked questions in a mathematics classroom is, 'What will I ever use this for?' Some teachers can give a good, convincing answer. Others hem and haw and stare at the floor. My favourite answer is, 'Algebra gives you *power*.' Algebra gives you the power to move on to bigger and better things in mathematics. Algebra gives you the power of knowing that you know something that your neighbour doesn't know. Algebra gives you the power to be able to help someone else with an algebra task or to explain to others these logical mathematical processes.

Algebra is a system of symbols and rules that is universally understood, no matter what the spoken language. Algebra provides a clear, methodical process that can be followed from beginning to end. It's an organisational tool that is most useful when followed with the appropriate rules. What power! Some people like algebra because it can be a form of puzzle-solving. You solve a puzzle by finding the value of a variable. You may prefer Sudoku or crosswords, but it wouldn't hurt to give algebra a chance, too.

About This Book

This book isn't like a mystery novel; you don't have to read it from beginning to end. In fact, you can peek at how it ends and not spoil the rest of the story.

I divide the book into some general topics — from the beginning nuts and bolts to the important tool of factoring to equations, applications and geometry. So you can dip into the book wherever you want, to find the information you need.

Throughout the book, I use many examples, each a bit different from the others, and each showing a different twist to the topic. The examples have explanations to aid your understanding. (What good is knowing the answer if you don't know how to get the right answer yourself?)

The vocabulary I use is mathematically correct *and* understandable. So whether you're listening to your teacher or talking to someone else about algebra, you'll be speaking the same language.

Along with the *how*, I show you the *why*. Sometimes remembering a process is easier if you understand why it works and don't just try to memorise a meaningless list of steps.

I don't use many conventions in this book, but you should be aware of the following:

- ✔ When I introduce a new term, I put that term in *italics* and define it nearby (often in parentheses).
- ✔ I express numbers or numerals either with the actual symbol, such as 8, or the written-out word: *eight*. Operations, such as +, are either shown as this symbol or written as *plus*. The choice of expression all depends on the situation — and on making it perfectly clear for you.

The *sidebars* (those little grey boxes) are interesting but not essential to your understanding of the text. If you're short on time, you can skip the sidebars. Of course, if you read them, I think you'll be entertained.

Foolish Assumptions

I don't assume that you're as crazy about maths as I am — and you may be even *more* excited about it than I am! I do assume, though, that you have a mission here — to brush up on your basic skills, improve your maths grade, or just have some fun. I also assume that you have some experience with algebra — for example, full exposure for a year or so.

You may remember the first time algebra came up in your maths class. I can distinctly remember my first algebra teacher, Miss McDonald, saying, 'This is an *n*.' My whole secure world of numbers was suddenly turned upside down. I hope your first reaction was better than mine.

Wherever you are in your maths journey, or what aspect you need to improve on, never fear. Help is here!

Icons Used in This Book

The little drawings in the margin of the book are there to draw your attention to specific text. Here are the icons I use in this book:

To make everything work out right, you have to follow the basic rules of algebra (or mathematics in general). You can't change or ignore them and arrive at the right answer. Whenever I give you an algebra rule, I mark it with this icon.

Paragraphs marked with the Remember icon help clarify a symbol or process. I may discuss the topic in another section of the book, or I may just remind you of a basic algebra rule that I discuss earlier.

The Tip icon isn't life-or-death important, but it generally can help make your life easier — at least your life in maths and algebra.

The Warning icon alerts you to something that can be particularly tricky. Errors crop up frequently when working with the processes or topics next to this icon, so I call special attention to the situation so you won't fall into the trap.

Where to Go From Here

If you want to refresh your basic skills or boost your confidence, start with Part I. If you're ready to jump into the guts of algebra, or looking for some factoring practice and need to pinpoint which method to use with what, go to Part II. Part III is for you if you're ready to solve equations; you can find just about any type you're ready to attack. Part IV is where the good stuff is — applications and geometry — things to do with all those good solutions. The list in Part V is usually what you'd look at after visiting one of the other parts, but why not start there? It's a fun place!

Studying more advanced maths and algebra can give you some logical exercises, and thinking logically can help you with all aspects of life — at school and afterwards.

The best *why* for studying algebra is just that it's beautiful. Yes, you read that right. Algebra is poetry, deep meaning and artistic expression. Just look and you'll find it. Also, don't forget that it gives you *power*.

Enjoy the adventure!

Part I
Reviewing the Basics

In this part ...

- Understand that maths can be a game — and that parents can play too.
- Work with addition and subtraction, and multiply with style and divide with ease.
- Get your head around negative numbers.
- Remember your fraction facts, and convert to and from percentages, decimals and fractions.
- Complete operations in the right order.

Chapter 1

Assembling Your Tools

In This Chapter

▶ Getting a refresher on whole numbers and fraction, decimals and percentages

▶ Starting to think about algebra

▶ Working out where geometry comes in

▶ Understanding the ways you can play with maths

▶ Focusing on how parents can help

*O*ne useful characteristic about numbers is that they're *conceptual,* which means that, in an important sense, they're all in your head. (This fact probably won't get you out of having to know about them, though — nice try!) For example, you can picture three of anything: three cats, three cricket balls, three cannibals, three planets. But just try to picture the concept of three all by itself, and you find it's impossible. Oh, sure, you can picture the numeral 3, but the *threeness* itself — much like love or beauty or honour — is beyond direct understanding. But when you understand the *concept* of three (or four, or a million), you have access to an incredibly powerful system for understanding the world: mathematics.

In this chapter, I run through some of the basics about working with whole and part numbers. I then provide an overview of some more advanced maths as I help you start to get your head around algebra and some aspects of geometry.

I then move on to some of the other tools you may like to assemble — including some tools for developing the right mindset towards maths and for connecting arguments and ideas using maths. I also provide some tools for parents helping their children work through their latest maths problem — especially valuable in the heat of the moment (you know, when the homework due tomorrow has just been discovered at the bottom of the schoolbag).

Working in with the Australian Curriculum

In recent years in Australia, the Australian Curriculum (AC) has been introduced, with the main objective of streamlining the curriculum for all subjects, across all states. In particular, the AC aims to ensure that students can move from state to state with minimal difference in subject content.

Schools can still use their discretion to order the topics within a subject and to organise different forms of assessment, but the core curriculum is to be the same across Australia. Through setting these consistent national standards, the AC aims to improve education outcomes for all Australian students. The website for the AC states that this then creates 'the base for future learning, growth and active participation in the Australian community'.

When focusing on maths in particular the website for the AC states:

> *The Australian Curriculum: Mathematics provides students with essential mathematical skills and knowledge in Number and Algebra, Measurement and Geometry, and Statistics and Probability. It develops the numeracy capabilities that all students need in their personal, work and civic life, and provides the fundamentals on which*

mathematical specialties and professional applications of mathematics are built.

(See www.australiancurriculum. edu.au for more information.)

But the Australian Curriculum: Mathematics doesn't just improve specialised maths skills. Over many, many years, mathematics has evolved and changed. The introduction of digital technology and significant changes in the types of calculators available provide new ways for students to develop their mathematical thinking and reasoning. Focusing on developing skills in key areas can enable students to improve their understanding, fluency and problem-solving abilities.

Both the initial construction and continued development of the Australian Curriculum: Mathematics has ensured that considered links exist between different subject areas, so students can develop an understanding of where mathematics fits within all subject areas and, therefore, everyday scenarios. Because of this, hopefully, students everywhere can learn to see not only the beauty, but also the relevance in mathematics and mathematical concepts.

Starting with the Basics

Where would mathematics and algebra be without numbers? A part of everyday life, numbers are the basic building blocks of algebra. Numbers give you a value to work with. Where would civilisation be today if not for numbers? Without numbers to figure the distances, slants, heights and directions, the pyramids would never have been built. Without numbers to figure out navigational points, the Vikings would never have left Scandinavia. Without numbers to examine distance in space, humankind could not have landed on the moon.

Even the simple tasks and the most common of circumstances require a knowledge of numbers. Suppose that your mum asked you to figure out the amount of petrol it takes for her to get from home to school and then on to work and back each day. You need a number for the total kilometres between your home, school and work and another number for the total kilometres your car can run on a litre of petrol.

It's sometimes really convenient to declare, 'I'm only going to look at whole-number answers', because whole numbers do not include fractions or negatives. You could easily end up with a fraction if you're working through a problem that involves a number of cars or people. Who wants half a car or, heaven forbid, a third of a person?

Whole numbers: Adding, subtracting, multiplying and dividing

Adding things up and taking them away are the two most fundamental skills in arithmetic. If you master these skills — just two sides of the same coin — you'll find the rest of this book much, much easier than it would be without them. I provide a refresher on all the basics in Chapter 2.

In Chapter 2, I also cover multiplication and division. You may have memories of reciting times tables in earlier years at school. I was terrified of my maths teacher at primary school, even though I was good at my times tables. Every Friday, he spent half the lesson marching up and down barking out 20 questions from the times tables. We learnt them soon enough, but teaching by intimidation is hardly the method I'd recommend.

Instead, in Chapter 2 I show you the times tables and give you some games to play to remember them. The times tables usually only go up to ten, so I also show you how to work with bigger numbers.

Dividing is exactly the opposite of multiplying: You take a number of things and split them into equal piles. Armies are split up into divisions. So are Aussie Rules football leagues. So '92 divided by four' just means 'split up 92 into 4 piles and tell me how big the piles are'. Or, 'split up 92 into piles of 4, and tell me how many piles there are'. In Chapter 2, I show you some games to help you remember your division sums up to $100 \div 10$, and then show you how to do division when you have bigger numbers. Again, you just need to split up piles.

In Chapter 3, I move on to negative and positive numbers.

Parts of the whole: Fractions, decimals and percentages

Seeing how whole numbers fit together is relatively easy, but then suddenly the evil maths guys start throwing fractions and percentages at you — and things aren't so intuitive. Fractions (at least, proper fractions) are just numbers that are smaller than whole numbers — they follow the same rules as regular numbers but sometimes need a bit of adjusting before you can apply them to everyday situations.

I have two main aims for Chapter 4: To show you that fractions, decimals and percentages are nothing like as fearsome as you may believe; and to show you that fractions, decimals and percentages are all different ways of writing the same thing — therefore, if you understand one of them, you can understand all of them.

I won't promise that you'll emerge from Chapter 4 deeply in love with fractions, but I hope I can help you make peace with fractions so you can work through the questions likely to come up in exams and in real life.

In Chapter 5 I take you through the idea of order of operations, so you know what you're supposed to be doing first (adding, multiplying, whatever), for whole numbers and for fractions.

Moving On to Algebra

You've probably heard the word *algebra* on many occasions, and no doubt you know that it has something to do with mathematics. But what exactly is algebra? What is it really used for?

Parts II and III answer these questions and more, providing the straight scoop on what it's good for, how algebra is used, and what tools you need to make it happen. In this chapter, you find some of the basics necessary to more easily find your way through the different topics in Parts II and III. I also point you toward these topics.

In a nutshell, algebra is a way of generalising arithmetic. Through the use of variables (letters representing numbers) and formulas or equations involving those variables, you solve problems. The problems may be in terms of practical applications, or they may be puzzles for the pure pleasure of the solving. Algebra uses positive and negative numbers, integers, fractions, operations, and symbols to analyse the relationships between values. It's a systematic study of numbers and their relationship, and it uses specific rules.

Speaking in Algebra

Algebra and symbols in algebra are like a foreign language. They all mean something and can be translated back and forth as needed. It's important to know the vocabulary in a foreign language; it's just as important in algebra.

Here's some important vocabulary for your journey into algebra:

- An *expression* is any combination of values and operations that can be used to show how things belong together and compare to one another. $2x^2 + 4x$ is an example of an expression. You see distributions over expressions in Chapter 6.

- A *term*, such as $4xy$, is a grouping together of one or more *factors* (variables and/or numbers). Multiplication is the only thing connecting the number with the variables. Addition and subtraction, on the other hand, separate terms from one another. For example, the expression $3xy + 5x - 6$ has three *terms*.

- An *equation* uses a sign to show a relationship — that two things are equal. By using an equation, tough problems can be reduced to easier problems and simpler answers. An example of an equation is $2x^2 + 4x = 7$. See the chapters in Part III for more information on equations.

- An *operation* is an action performed upon one or two numbers or terms to produce a resulting number. Operations are addition, subtraction, multiplication, division, square roots, and so on. See Chapter 2 for more on operations.

- A *variable* is a letter representing some unknown; a variable always represents a number, but it *varies* until it's written in an equation or inequality. (An *inequality* is a comparison of two values.) Then the fate of the variable is set — it can be solved for, and its value becomes the solution of the equation. By convention, mathematicians usually assign letters at the end of the alphabet to be variables (such as x, y and z).

- A *constant* is a value or number that never changes in an equation — it's constantly the same. Five is a constant because it is what it is. A variable can be a constant if it is assigned a definite value. Usually, a variable representing a constant is one of the first letters in the alphabet. In the equation $ax^2 + bx + c = 0$, a, b and c are constants and the *x* is the variable. The value of *x* depends on what *a*, *b* and *c* are assigned to be.

✔ An *exponent* is a small number written slightly above and to the right of a variable or number, such as the 2 in the expression 3^2. It's used to show repeated multiplication. An exponent is also called the *power* of the value. For more on exponents, see Chapter 6.

Taking aim at algebra operations

In algebra today, a variable represents the unknown. Before the use of symbols caught on, problems were written out in long, wordy expressions. Actually, using letters, signs and operations was a huge breakthrough. First, a few operations were used, and then algebra became fully symbolic. Nowadays, you may see some words alongside the operations to explain and help you understand, like having subtitles in a movie.

By doing what early mathematicians did — letting a variable represent a value, then throwing in some operations (addition, subtraction, multiplication, and division), and then using some specific rules that have been established over the years — you have a solid, organised system for simplifying, solving, comparing or confirming an equation. That's what algebra is all about: That's what algebra's good for.

What About Geometry?

Studying geometry is sort of a Dr Jekyll-and-Mr Hyde thing. You have the ordinary geometry of shapes (the Dr Jekyll part) and the strange world of geometry proofs (the Mr Hyde part).

Every day, you see various shapes all around you (triangles, rectangles, boxes, circles, balls and so on), and you're probably already familiar with some of their properties: area, perimeter, and volume, for example. In Chapters 15 and 16, you discover much more about these basic properties and then explore more advanced geometric ideas about shapes.

Geometry proofs are an entirely different sort of animal. They involve shapes, but instead of doing something straightforward like calculating the area of a shape, you have to come up with a mathematical argument that proves something about a shape. This process requires not only mathematical skills but verbal skills and logical deduction skills as well, and for this reason, proofs trip up many, many students. If you're one of these people and have already started singing the geometry-proof blues, you might even describe geometry proofs — like Mr Hyde — as monstrous. But I'm confident that, with the help of the chapters in Part IV, you'll have no trouble taming some of them.

Playing with Maths

Ideas are things you can play with, and maths is about ideas. Being able to approach maths as a game of ideas is another important tool for your toolbox.

When children make up stories, parents and teachers are usually delighted to listen. Children are brought up with books and stories, and watching them practise writing their own can be exciting. When children make up games, build things or start playing instruments, parents and teachers can see the connection to playing.

But with maths, people's vision can get clouded. It's easy to be concerned with *right* and *wrong* in math, while losing sight of the importance of playing with ideas. Yet playing is just as possible and just as useful in maths as it is in other areas of children's lives. These sections look at some examples of maths play, as they may happen in classrooms and at home.

Experimenting with symbols

Mathematical symbols have meaning, and you should really try to keep that meaning in mind as you work. Okay, I realise that it doesn't sound very playful, so stick with me here.

For example, think back to primary school and imagine a class full of second graders discussing how many muffins are in the partially filled muffin tin in Figure 1-1. Some of these second graders will count the muffins one by one. Others will see more sophisticated relationships. The teacher may help students record their thinking using arithmetic. For example:

- $3 + 3 + 1$
- $12 - 3 - 2$
- $\frac{1}{2}$ of $12 + 1$
- $2 + 2 + 3$
- $3 \times 3 - 2$

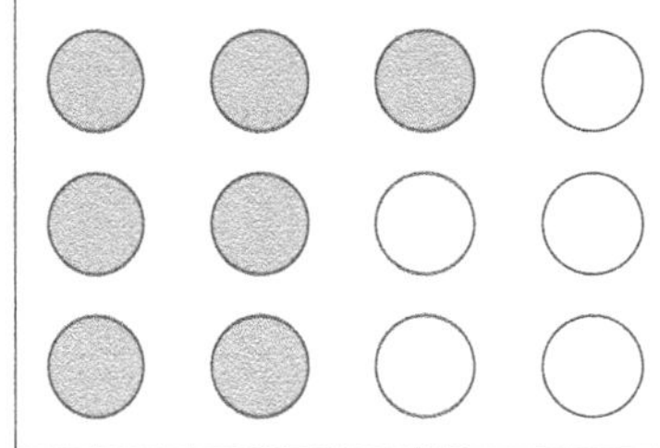

Can you see how each of these expressions correctly counts the muffins? What could the child who wrote each expression have been thinking?

When you (or a second grader) are working on a task such as this one, you're playing with maths. You're imagining new relationships, trying something without fear of getting it wrong, and seeing a small corner of the world in new ways. Children do all these things when they play.

No *one answer* is involved when you ask, 'How can you count these muffins?' All of the ways of counting come up with seven muffins. That leaves you free to play with the ideas and see whether you can see what someone else sees in this situation.

Building models

Building is a form of play. Building with blocks, building sand castles, building a catapult to launch a cricket ball across the backyard (and hopefully not breaking any of next door's windows), and building a mathematical model are all different forms of the same playful human instinct to create.

When children make predictions, they're building mathematical models. For example, my son's bedtime went from 7.30 to 8.00 when he turned 8 years old. He took this to be a rule and began predicting future bedtimes at future ages, assuming that his bedtime would advance by a half hour each year. He extrapolated in both directions to determine that he will go to bed at 2.30 in the morning when he turns 21 and that he must have gone to bed at 6.00 when he was 4 years old. He built a model of the relationship between his bedtime and his age, and he played with that relationship.

Similar things happen in maths classrooms. Students identify patterns, they make predictions based on these patterns, and they check their predictions against reality.

Arguing is heaps of fun

An *argument* in maths isn't really the same as an argument you may have with your siblings or friends in the back seat of the car. The back seat argument can devolve quickly into name-calling, spitballs and fisticuffs if your parents don't intervene. This behaviour shouldn't happen in a mathematical argument. What these two kinds of arguments have in common, however, is disagreement. In maths, the disagreement is usually around whether something is true. And constructing viable arguments and critiquing the reasoning of others is at the heart of maths as a discipline.

Although many people think maths is about computing with numbers and variables, it's really about making particular kinds of arguments. Mathematicians base their arguments on things they already know are true and on the rules of logic. If something is true in maths, you can know it in a much more certain and timeless way than in any other subject.

You can practise making arguments by asking one simple question on a regular basis: 'How do I know?' For example, if you know your little sister ate four pistachios because she has eight pistachio shells on her plate, ask yourself how you know eight shells mean four pistachios. Put the argument into words. Make a habit of asking yourself how you know to get into the habit of building mathematical arguments.

If you have ever tried to explain to a child (perhaps a younger sibling or cousin) exactly why he can't do something, you have likely noticed that young children are very good at critiquing the reasoning of others. Even toddlers can be like little lawyers when lollies or some other privilege or liberty are at stake. While this can be tremendously frustrating, it's a highly useful mathematical technique. Use it when you can (but still expect your parents to continue to lay down the law when they need to).

Connecting ideas

Remembering a set of disconnected facts is more difficult than remembering a story. The typical human brain is very good at telling and remembering stories, while also very bad at retaining lists. A major difference between a list and a story is connections.

Fortunately (and not coincidentally), maths is rich with connections. *Connections* basically are relationships between ideas. But maths is often taught in a way that obscures these connections. When teachers insist on you memorising both addition and subtraction facts (which is different from insisting that students be able to produce these facts quickly), they obscure the connections between addition and subtraction, for example. If you know that $8 + 4 = 12$ and you know the connections between addition and subtraction, you can quickly produce $4 + 8$, $12 - 4$ and $12 - 8$ because they're all connected examples of one relationship.

Mathematics can connect what seem to be very different situations. When you build a mathematical model of a situation, you strip away details. For example, when you have three red apples and four green apples and you write $3 + 4$, you strip away the fact that this is an idea about apples. Then, when you count four spaces to the right on the number line, starting at 3, and you write $3 + 4$ again, you have a connection. This relationship — that $3 + 4 = 7$ — is true of a wide variety of contexts.

The following two questions are common in algebra and geometry classes:

- If n people are on a basketball team and each one gives the others one high five, how many high fives are given altogether?
- How many diagonals can you draw in a regular polygon with n sides?

In both cases, the answer is the same: $\frac{n(n-1)}{2}$. After you have enough experience with these types of problems, you can start to have hunches about which kinds of mathematical models are likely to be useful for different situations. Through repeated experiences with modelling, you can get better at noticing the structure of a problem situation.

In addition to thinking of algebra as *generalised arithmetic* (this means that algebra answers questions about *all* numbers, not just the numbers in a particular computation), you can think of algebra as an efficient way of getting things done. Algebra can capture the regularity in repeated reasoning. In order to capture it, you need to look for connections.

A simpler example of regularity in repeated reasoning occurs when you move from counting to solve problems such as $9 + 2$, $9 + 3$ and $9 + 4$ to having strategies for knowing the sum of 9 and any one-digit number. Perhaps you've already noticed that these sums come out as 10 plus a number one less than the original number. That is $9 + 2 = 11$, and 11 is $10 + 1$. Even if you haven't memorised all of your single-digit addition facts, you may have noticed the regularity in the repeated reasoning about sums involving 9 — and similar patterns in sums involving other numbers.

What Parents Can Do to Help

As parents see their children struggling with their maths homework, or struggling to improve their maths marks, many want to offer some help and guidance. The following sections show you how to provide this help — without taking over.

Focusing on asking questions

Asking questions is a huge part of learning. Many maths teachers now incorporate asking questions into their instruction, but that may be different from how you imagine based on your own experience as a math student.

In a typical maths class, students should be asking all different types of questions of themselves and of others. This section takes a closer look at these types of questions.

Asking 'Why?' and 'How do you know?'

Questions, such as 'why' and 'how do you know' (or 'how do *I* know'), are very similar. They force you to build an argument.

One common claim about maths in the early years is that students need to *just know* their number facts, such as all the sums of two one-digit numbers $(3 + 8, 7 + 5,$ and so on). Students certainly do need to be able to produce single-digit sums quickly, but they can also think more deeply about the sum.

You may not have stopped to consider why $8 + 4 = 12$ for quite some time. In fact, asking the question, 'Why is $8 + 4 = 12$?' may seem totally silly. You just know that it is.

But a slightly different version of the question may seem more compelling. How do you *know* that $8 + 4 = 12$? Or consider this version: If you couldn't remember what $8 + 4$ was for a moment, how could you figure it out?

Maybe you could count it out. Start at 8 and then count four more — 9, 10, 11 and 12. That counting ends on 12, so $8 + 4 = 12$. Although this strategy is a lot less practical for something such as $9 + 9$, and it doesn't transfer well at all to multi-digit addition, it's a viable strategy for explaining why 8 and 4 make 12.

Here's another way to know this number fact — one that is somewhere between counting and memorising yet is powerful enough to work with many different kinds of numbers. Ten is an important number (the system for naming numbers is based on ten). If you know that $8 + 2 = 10$ and you also know that 4 is two more than 2, then $8 + 4$ is two more than $8 + 2$, so it's two more than 10, which is 12.

When students relate number facts to each other and use ideas in addition to recalling facts, they're making use of a maths structure.

Asking 'What if?'

This question is at the heart of nearly every historical advancement in mathematics and should be at the heart of every student's learning. Think about negative numbers, for example. Nearly everyone has the experience of being in second grade and listening to the teacher explain how to subtract $35 - 19$ using the standard algorithm. 'You can't take 9 from 5,' the teacher says, referring to the units place, 'so you need to borrow from the tens place.'

But what if you *could* take 9 from 5? What if $5 - 9$ *did* have an answer? Second graders sometimes ask similar important questions, and the answers to these questions involve negative numbers. A whole new kind of number comes from asking 'What if?'

Similarly, fractions come from asking, 'What if you could share 5 things equally among 3 people?' and complex numbers come from asking, 'What if you could find a square root of a negative number?'

Asking 'Is it good enough?'

Maths is often seen as a precise discipline, and students attend to precision when they ask themselves, 'Is it good enough?' This question is about numbers, but it's also about communication. At a basic level, precision means making judgements about measurements and calculations by answering questions such as these:

- When is it okay to estimate?
- When is it okay to use 3.14 for π?
- Should I round this number? To what place?
- Is $\frac{1}{3}$ really equal to 0.33?

The answers to these questions relate to being precise about the use of numbers, which is important in mathematics. Just as important, though — and often overlooked — is attending to precision in other areas of doing and communicating mathematics.

Being careful with units is an important part of doing mathematics, and it's part of *attending to precision*. Any time you express a quantity, a unit goes along with it. Two plus two equals four because two *things* and two more *things* total four *things*.

Paying attention to units is important when students are learning about multiplication (see Chapter 2 for more about units and multiplication) because 3×4 means 3 *groups of* 4. The 3 and the 4 refer to different units. You can't multiply three eggs by four eggs and get anything meaningful. But you can multiply three cartons by 4 eggs per carton to get 12 eggs all together.

Paying attention to units is important when students are studying fractions because the algorithm for adding fractions depends on units (see Chapter 4): $\frac{1}{4} + \frac{2}{4} = \frac{3}{4}$ because *fourths* are units. You have one-fourth and two more fourths for a total of three-fourths.

A final example of attending to precision is in the use of language. Mathematicians use words very precisely, and it begins with precise definitions. *A circle is a round shape* isn't a precise definition — it isn't good enough to sort out circles from all things that aren't circles. An oval is round but not a circle, for example. A precise definition for a circle is *the set of points in a plane that are the same distance from a common point, called the centre*. In a maths class, the need for precise definitions should come from exploring ideas and arguing about the meaning of words. Questions such as 'Is a square a rectangle?,' 'Is 0 even?,' 'Is 1 prime?,' and 'Is $y = 1$ a linear function?' all depend on precise definitions of the relevant words — *rectangle, even, prime* and linear.

Attention to the precision of a definition doesn't need to wait for late primary or high school, though. Young children are ready to play with definitions even before they attend school. 'Is a hot dog a sandwich?' is a lovely question accessible to young children that helps them attend to the precision of the meaning of *sandwich*. Is your definition of sandwich good enough to determine whether a hot dog is one? Making lunch can be a chance to work on mathematical practices!

Asking 'Does this make sense?'

Students need to make sense of problems when they do maths. The meaning of *a problem* varies widely from kindergarten to high school, but making sense of problems is important throughout. 'Does this make sense?' is an important question for students to ask themselves as they work.

Grade One's may meet a problem such as this: Elle and Josh bought some apples, but Josh ate five while Elle was at work. Now there are seven left. How many did they buy?

A student who tries to solve that problem *without* making sense of it may search for so-called keywords, find the word *left*, remember that it means to subtract, and subtract five from seven to get two. For a first grader, making sense of a problem means understanding what is going on, which may involve drawing a picture or telling a story that relates the problem to her own life. It does not mean just translating the problem into mathematical symbols. A first grader needs to ask herself, 'Does it make sense to subtract these numbers?'

A student in Year 9, by contrast, may have a problem based purely in mathematical relationships, such as 'Does doubling the *x*-value in a linear function always double the *y*-value? Does it ever?' Making sense of this problem for a Year 9 student requires drawing on more vocabulary (*linear, function* and so on) than a first grader will likely require, and it involves a good deal more sophistication. But the idea is the same: Making sense of problems is very different from knowing what value to substitute and computing correctly.

Asking whether something makes sense is part of persevering. When you ask the question, the answer may be *no*, and that may require trying again. At all grade levels, when students persevere in solving problems, they're willing to accept the fact that something they try may not work. They're ready to start over again and to spend time trying to figure out things. A classroom (and home) climate that values working hard and learning from mistakes is helpful in fostering perseverance in problem solving. A climate that only values right answers is less likely to encourage students to persevere.

Asking 'What's going on here?'

Structure refers to how things are put together in maths and to how things relate to each other. Structure reveals what is going on beneath the surface. You can make a habit of looking for structure by asking, 'What's going on here?'

If you have \$12.00 and I have \$3.00, you could say that you have \$9.00 more than me. That is making use of the *additive structure* of numbers — $3 + 9 = 12$. But you could also say that you have four times as much as I do. In that case, you're using the multiplicative structure of numbers — $4 \times 3 = 12$. You can ask, 'What is going on with this comparison?' and 'Does it make more sense to compare by multiplying and dividing, or by adding and subtracting in this situation?' (Chapter 2 discusses the additive and multiplicative structures in greater depth.)

When students sort shapes, they can pay attention to what the shapes look like or they can pay attention to the mathematical structure of the shapes, such as the number of sides, relationships among the sides or the measures of their angles. This transition from noticing that rectangles look like doors, for example, to noticing that rectangles have four right angles is a long process in school that begins in kindergarten. The question 'What is going on with these rectangles that makes them rectangles?' is important here.

Look at the sequence of shapes in Figure 1-2. Each shape is made of one or more square tiles. If the shapes were to continue growing in the same way these do, how many tiles would be in the fifth shape? How many would be in the tenth shape? How many would be in the 100th shape?

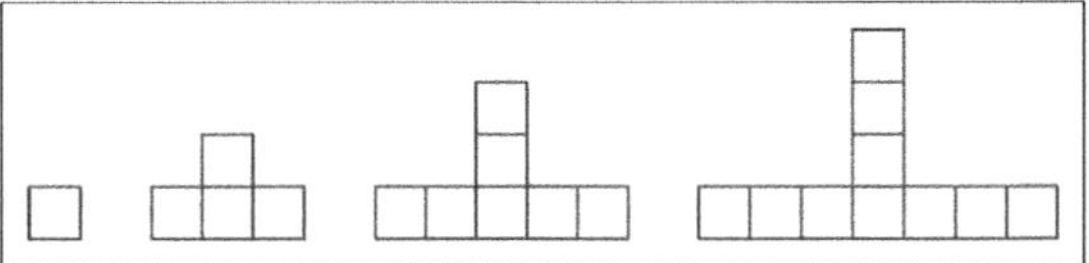

Figure 1-2: From left to right, the first through fourth shapes in a sequence.

Using the structure of these shapes, you may say that each shape consists of a tile in the centre (shaded in Figure 1-3), surrounded by three arms, each with the same number of tiles. If the sequence number of the shape is n, the length of each of these arms is $n - 1$, so the total number of tiles is $3(n - 1) + 1$.

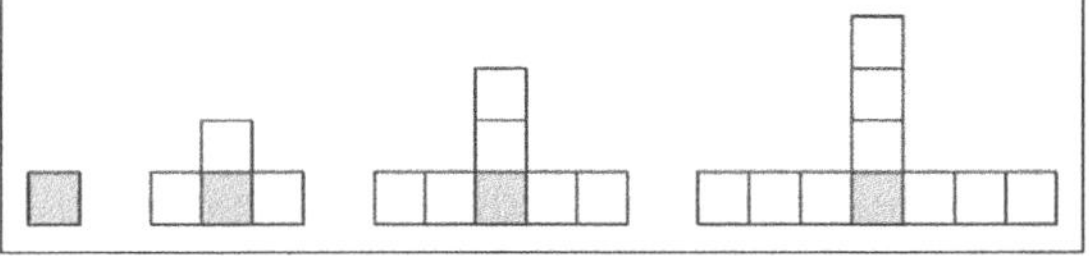

Figure 1-3: One way to see the structure of the shapes.

Other people may see different structures. The point is that seeing the structure of these shapes allows you to calculate the number of tiles in any of these shapes. The 100th shape has 298 tiles, for example.

Looking for and making use of structure is what students do when they do mental maths. Something as simple as multiplying 4×10 by thinking about it as *four tens* or *forty* is an early place-value example of structure.

A final example of looking for structure is noticing the big picture of what algebraic symbols are telling you. If you know that n is a whole number, for example, then $2n$ is an even number and $2n + 1$ is an odd number. What's more, the expression $2n + 1$ represents all odd numbers. This is the power of looking for and making use of structure — representing infinitely many things in a single short expression.

Helping your child with homework (without doing the work yourself)

Whatever the standards may be — based on the Australian Curriculum or anything else — most children will be frustrated with a homework assignment from time to time. The advice to you, the parent, doesn't change.

What may have changed since you went to school is the ways children are expected to work on their homework. Teachers may ask their students to practise something they worked on in class, but it may look unfamiliar to you. Productive ways of helping children with their homework are the same, though.

Researchers have found that the human mind is surprisingly lazy. If it can get something done without thinking hard, it will. This research is no surprise to someone who has witnessed a typical teenager eager to get through her maths homework so that she can go and watch TV. 'Is this right?' she asks, hopefully. 'Is it right now? What about this?' This routine can quickly devolve into a guessing game rather than a chance to learn anything useful.

An important function of homework is what teachers call *formative assessment*, which means learning what students understand while they're still studying it. A classroom full of correct homework papers can signal to a teacher that everyone understands and it's time to move on. Now imagine if those correct homework papers were the result of parents (or tutors or older siblings) walking these students through the steps of the homework problems. It could mislead the teacher. It's much better to write a note explaining what the student is struggling with (or to let your child submit some incorrect or incomplete problems) than to let the teacher think that your child understands something that he doesn't.

Don't do the homework for your child. Help your child clarify his thinking and identify what he knows and doesn't know. Monitor the difficulty level to make sure your child has interesting and challenging work, but not work that is far beyond his present abilities. Keep in touch with his teacher if things are out of balance so that you can work together for your child's benefit.

In the heat of the moment, though, you can easily lose sight of the big picture. So here are three simple tips for productive involvement with maths homework and for getting students to think instead of guess and so to learn something rather than just stumble onto a right answer:

- **Ask 'How do you know?'** As covered in the preceding section, this question forces students to think about their own thinking, which is an important part of making that thinking better. Ask this question frequently. Question right and wrong answers. Also keep in mind that this question has many variations. 'How do you know this is right?,' 'How did you know that was wrong?,' 'How did you know to do that?,' and so on.

- **Wait for a response.** Ten or 15 seconds seems like a long time to sit silently when you know the answer, but it's not long at all to the person trying to figure the answer out. What goes on in the silent time between asking a question and getting an answer is thinking. One of the most important findings in educational research is that increased *wait time* — the time between a teacher asking a question and the next time someone speaks — is strongly associated with increased learning. When teachers give their students more time to think about their questions, students learn more. It's true at home, too.

- **Share a strategy.** After your child explains her thinking, talk about your own. Don't tell her how she needs to do something; just tell her in the spirit of sharing your own ideas. It's like being at the dinner table. When you want your child to share something that happened during the day, you model it by sharing your own stories from the day. It's the same with thinking. If you want your child to engage with maths homework, you can model that behaviour by talking about how *you* think about these problems.

Getting the inside scoop: Teachers' views on homework

Knowing something about teachers' reasons for assigning homework can help you to know how to support your child at homework time. The issue of homework can be more complicated than you might expect, and teachers assign homework for many reasons. Here is a partial list:

✔ To practise skills learned in class

✔ To introduce new ideas that weren't covered in class due to lack of time

✔ To set the stage for the next day's lesson

✔ To teach responsibility

✔ To review things from previous grades that will be needed in the current grade

✔ To conform to school or state guidelines

✔ To conform to the expectations of parents and students

✔ To compel students to think about maths outside of the classroom

✔ To assess what students know in order to assign grades or to make decisions for future lessons

Most teachers can probably point to two or more of these reasons as their own. Many teachers can probably add another reason or two to the list.

Most teachers prefer to view themselves as partners with parents with the common interest of their students' success. To that end, you can ask at the beginning of the year (or anytime, really) what your child's teacher has as goals for homework and how she would like you to help out when your child gets stuck.

Most teachers and parents agree that a child's emotional and physical wellbeing needs to take precedence over any particular homework assignment. If your child occasionally struggles with homework, you may be able to let it go for a night or two as long as you make a plan with your child and stay in touch with the teacher. Write a note to the teacher explaining the situation and help your child catch up on necessary work over a few days or the weekend.

If you notice a pattern where your child is consistently struggling with homework, though, you'll want to sit down with the teacher to find out her purposes for homework and to develop strategies to meet those purposes.

Becoming unstuck: What to do

A few of the more stressful parenting moments include when time is tight and when your child is struggling with a homework assignment. Instead of cursing the heavens, Pythagoras and the originators of homework, you need a strategy. Here is a good one, in three parts:

✔ **Listen:** Ask your child to tell you what she is supposed to do. If something needs to be done a particular way, listen to what it is. Ask what she knows and doesn't know and then listen to the response. Don't try to tell her what to do or show her the way that you remember from school. Just listen.

✔ **Help:** After you have listened, decide how you can help. Sometimes you may have to show your child the way you learned to do something. Sometimes you may have to look something up online together. Sometimes you may have to phone a friend — your child's friend who may have an idea about what to do or your friend who may have some expertise. Don't assume that the way you learned is the best way (nor that it is the worst). An alternate way of working through something may help a child understand the meaning of what she's doing.

✔ **Let it go:** You and your child shouldn't be expected to spend your evenings engaged in tears and combat in the name of multiplication facts. As a parent, model good study habits by establishing that some time will be spent doing what is assigned. You also should model good mental health habits by setting limits on that time. When your child is no longer making productive progress, let this homework go. You can write a note to the teacher explaining the decision that you have made and get on with the rest of your evening.

If getting stuck is a rare or occasional thing, letting it go is a good strategy to practise. If your child is regularly getting stuck on her homework and you find yourself letting it go on a regular basis, you probably need to have a conversation with your child's teacher so that you each understand the demands being placed on your child. Frequent homework struggles can be a sign that your child needs something different from what she's getting at school. In such a case, you and your child's teacher should figure out how to work together to get her what she needs.

Chapter 2

Working with Whole Numbers

In This Chapter

▶ Reviewing addition

▶ Understanding subtraction

▶ Viewing multiplication as a fast way to do repeated addition

▶ Getting clear on division

*W*hen most people think of maths, the first thing that comes to mind is four little (or not-so-little) words: Addition, subtraction, multiplication and division. I call these operations the main four basic operations.

In this chapter, I introduce you (or reintroduce you) to these little gems. Although I assume you're already familiar with these four operations, this chapter reviews them, taking you from what you may have missed to what you need to succeed as you move onward and upward in maths.

Adding Things Up

Addition is the first operation you find out about, and it's almost everybody's favourite. It's simple, friendly and straightforward. No matter how much you worry about maths, you've probably never lost a minute of sleep over addition. Addition is all about bringing things together, which is a positive goal. For example, suppose you and I are standing in line to buy tickets for a movie. I have $30 and you have only $10. I could lord it over you and make you feel crummy that I can go to the movies and you can't. Or instead, you and I can join forces, adding together my $30 and your $10 to make $40. Now, not only can we both see the movie, but we may even be able to buy some popcorn, too.

Addition uses only one sign — the plus sign (+): Your equation may read $2 + 3 = 5$, $12 + 2 = 14$ or $27 + 44 = 71$, but the plus sign always means the same thing.

When you add two numbers together, those two numbers are called *addends*, and the result is called the *sum*. So in the first example, the addends are 2 and 3, and the sum is 5.

In line: Adding larger numbers in columns

When you want to add larger numbers, stack them on top of each other so that the ones digits line up in a column, the tens digits line up in another column, and so on. Then add column by column, starting from the ones column on the right. Not surprisingly, this method is called *column addition*. Here's how you add $55 + 31 + 12$. First add the ones column:

$$\begin{array}{r} 55 \\ 31 \\ +12 \\ \hline 8 \end{array}$$

Next, move to the tens column:

$$\begin{array}{r} 55 \\ 31 \\ +12 \\ \hline 98 \end{array}$$

This problem shows you that $55 + 31 + 12 = 98$.

Carry on: Dealing with two-digit answers

Sometimes when you're adding a column, the sum is a two-digit number. In that case, you need to write down the ones digit of that number and carry the tens digit over to the next column to the left — that is, write this digit above the column so you can add it with the rest of the numbers in that column. For example, suppose you want to add $376 + 49 + 18$. In the ones column, $6 + 9 + 8 = 23$, so write down the 3 and carry the 2 over to the top of the tens column:

$$\begin{array}{r} 2 \\ 376 \\ 49 \\ +\ 18 \\ \hline 3 \end{array}$$

Now continue by adding the tens column. In this column, $2 + 7 + 4 + 1 = 14$, so write down the 4 and carry the 1 over to the top of the hundreds column:

$$\begin{array}{r} 12 \\ 376 \\ 49 \\ +\ 18 \\ \hline 43 \end{array}$$

Continue adding in the hundreds column:

$$\begin{array}{r} 12 \\ 376 \\ 49 \\ +\ 18 \\ \hline 443 \end{array}$$

This problem shows you that $376 + 49 + 18 = 443$.

This process applies no matter how large the numbers themselves become. The following example shows the steps for adding two five-digit numbers — $12{,}495 + 14{,}821$.

$$\begin{array}{r} ^{1\ 1} \\ 12{,}495 \\ +14{,}821 \\ \hline 27{,}316 \end{array}$$

Here's the problem broken down:

1. **Add the first column.**

   In this example, $5 + 1 = 6$, so you write **6** in the answer space.

2. **Add the second column.**

   Here, $9 + 2 = 11$, so you write the **1** in the answer space and carry the 1 above the 4.

3. Continue adding the columns, from right to left.

So, in the next column, $1 + 4 + 8 = 13$, so you write the **3** in the answer space and carry the 1 above the 2.

Moving to the next column, $1 + 2 + 4 = 7$, so you write **7** in the answer space. Finally, $1 + 1 = 2$, so you write **2** in the answer space.

So $12,495 + 14,821 = 27,316$.

This process also applies if you have more than two or three numbers to add, as the following example shows. Remember you can add zeros to fill the empty spaces if that helps keep the digits aligned.

$$
\begin{array}{r}
{\scriptstyle 1\ \ 11} \\
23,562 \\
65,321 \\
00,567 \\
+\ 00,015 \\
\hline
89,465
\end{array}
$$

Here's how you work through the problem:

1. **Add the first column.**

   In this example, $2 + 1 + 7 + 5 = 15$, so you write the **5** in the answer space and carry the 1 above the 6.

2. **Add the second column.**

   Here, $1 + 6 + 2 + 6 + 1 = 16$, so you write the **6** in the answer space and carry the 1 above the 5.

3. **Continue adding the columns, from right to left.**

   So, in the next column, $1 + 5 + 3 + 5 + 0 = 14$, so you write the **4** in the answer space and carry the 1 above the 3.

   Next, $1 + 3 + 5 + 0 + 0 = 9$, so you write **9** in the answer space.

   Finally, $2 + 6 + 0 + 0 = 8$, so you write **8** in the answer space.

Using grid paper can help you line up the digits correctly and carry them into the correct position.

Take It Away: Subtracting

Subtraction is usually the second operation you discover, and it's not much harder than addition. Still, there's something negative about subtraction — it's all about who has more and who has less. Suppose you and I have been running on treadmills at the gym. I'm happy because I ran 3 kilometres, but then you start bragging that you ran 10 kilometres. You subtract and tell me that I should be very impressed that you ran 7 kilometres farther than I did. (But with an attitude like that, don't be surprised if you come back from the showers to find your running shoes filled with liquid soap!)

As with addition, subtraction has only one sign: the minus sign (–). You end up with equations such as $4 - 1 = 3$, and $14 - 13 = 1$, and $93 - 74 = 19$.

When you subtract one number from another, the result is called the *difference*. This term makes sense when you think about it: When you subtract, you find the difference between a higher number and a lower one.

In subtraction, the first number is called the *minuend* and the second number is called the *subtrahend*. But almost nobody ever remembers which is which, so when I talk about subtraction, I prefer to say *the first number* and *the second number*.

One of the first facts you probably heard about subtraction is that you can't take away more than you start with. In that case, the second number can't be larger than the first. And if the two numbers are the same, the result is always 0. For example, $3 - 3 = 0$; $11 - 11 = 0$; and $1,776 - 1,776 = 0$. Later someone breaks the news that you *can* take away more than you have. When you do, though, you need to place a minus sign in front of the difference to show that you have a negative number, a number below 0:

$$4 - 5 = -1$$
$$10 - 13 = -3$$
$$88 - 99 = -11$$

When subtracting a larger number from a smaller number, remember the words *switch* and *negate*: You *switch* the order of the two numbers and do the subtraction as you normally would, but at the end, you *negate* the result by attaching a minus sign. For example, to find $10 - 13$, you switch the order of these two numbers, giving you $13 - 10$, which equals 3; then you negate this result to get –3. That's why $10 - 13 = -3$.

The minus sign does double duty, so don't get confused. When you stick a minus sign between two numbers, it means the first number minus the second number. But when you attach it to the front of a number, it means that this number is a negative number.

I also go into more detail on negative numbers and the main four operations in Chapter 3.

Columns and stacks: Subtracting larger numbers

To subtract larger numbers, stack one on top of the other as you do with addition. (For subtraction, however, don't stack more than two numbers — put the larger number on top and the smaller one underneath it.) For example, suppose you want to subtract 386 – 54. To start, stack the two numbers and begin subtracting in the ones column: 6 – 4 = 2:

$$\begin{array}{r} 386 \\ -54 \\ \hline 2 \end{array}$$

Next, move to the tens column and subtract 8 – 5 to get 3:

$$\begin{array}{r} 386 \\ -54 \\ \hline 32 \end{array}$$

Finally, move to the hundreds column. This time, 3 – 0 = 3:

$$\begin{array}{r} 386 \\ -54 \\ \hline 332 \end{array}$$

This problem shows you that 386 – 54 = 332.

Can you spare a ten? Borrowing to subtract

Sometimes the top digit in a column is smaller than the bottom digit in that column. In that case, you need to borrow from the next column to the left. Borrowing is a two-step process:

1. **Subtract 1 from the top number in the column directly to the left.**

   Cross out the number you're borrowing from, subtract 1, and write the answer above the number you crossed out.

2. **Add 10 to the top number in the column you were working in.**

For example, suppose you want to subtract 386 − 94. The first step is to subtract 4 from 6 in the ones column, which gives you 2:

$$
\begin{array}{r}
386 \\
-94 \\
\hline
2
\end{array}
$$

When you move to the tens column, however, you find that you need to subtract 8 − 9. Because 8 is smaller than 9, you need to borrow from the hundreds column. First, cross out the 3 and replace it with a 2, because 3 − 1 = 2:

$$
\begin{array}{r}
2 \\
\cancel{3}86 \\
-94 \\
\hline
2
\end{array}
$$

Next, place a 1 in front of the 8, changing it to an 18, because 8 + 10 = 18:

$$
\begin{array}{r}
2 \\
\cancel{3}186 \\
-94 \\
\hline
2
\end{array}
$$

Now you can subtract in the tens column: 18 − 9 = 9:

$$
\begin{array}{r}
2186 \\
-94 \\
\hline
92
\end{array}
$$

The final step is simple: $2 - 0 = 2$:

```
 2186
  -94
 ----
  292
```

Therefore, $386 - 94 = 292$.

In some cases, the column directly to the left may not have anything to lend. Suppose, for instance, that you want to subtract $1{,}002 - 398$. Beginning in the ones column, you find that you need to subtract $2 - 8$. Because 2 is smaller than 8, you need to borrow from the next column to the left. But the digit in the tens column is a 0, so you can't borrow from there because the cupboard is bare, so to speak:

```
 1002
 -398
 ----
```

When borrowing from the next column isn't an option, you need to borrow from the nearest non-zero column to the left.

In this example, the column you need to borrow from is the thousands column. First, cross out the 1 and replace it with a 0. Then place a 1 in front of the 0 in the hundreds column:

```
  0
 + 10 0 2
   -3 9 8
```

Now cross out the 10 and replace it with a 9. Place a 1 in front of the 0 in the tens column:

```
  0  9
 + 10 102
   -3  98
```

Finally, cross out the 10 in the tens column and replace it with a 9. Then place a 1 in front of the 2:

```
  0  9  9
 + 10 10 12
   -3  9  8
```

At last, you can begin subtracting in the ones column: $12 - 8 = 4$:

$$
\begin{array}{r}
0 \quad 9 \quad 9 \\
+ \ \cancel{10} \ \cancel{10} \ \mathbf{12} \\
-3 \quad 9 \quad \mathbf{8} \\
\hline
\mathbf{4}
\end{array}
$$

Then subtract in the tens column: $9 - 9 = 0$:

$$
\begin{array}{r}
0 \quad 9 \quad \mathbf{9} \\
+ \ \cancel{10} \ \cancel{10} \ 12 \\
-3 \quad \mathbf{9} \quad 8 \\
\hline
\mathbf{0} \quad 4
\end{array}
$$

Then subtract in the hundreds column: $9 - 3 = 9$:

$$
\begin{array}{r}
0 \quad 9 \quad \mathbf{9} \\
+ \ \cancel{10} \ 10 \ 12 \\
-3 \quad \mathbf{9} \quad 8 \\
\hline
\mathbf{6} \quad \mathbf{0} \quad 4
\end{array}
$$

Because nothing is left in the thousands column, you don't need to subtract anything else. Therefore, $1{,}002 - 398 = 604$.

This process applies no matter how large the numbers themselves become, as the following example shows. (Remember that you can add zeros to fill in spaces at the front of the number to help keep everything aligned.)

In this example, you need to work out $1{,}609{,}452 - 413{,}651$. Note that a zero has been added to the number being subtracted, so that there are the same number of digits in each number.

$$
\begin{array}{r}
{}^{510} \ {}^{8} \ {}^{14} \\
1{,}\cancel{609}{,}452 \\
-0{,}413{,}651 \\
\hline
1{,}195{,}801
\end{array}
$$

Here's how to break down the problem:

1. **Subtract the numbers in the first column.**

   In this example, $2 - 1 = 1$, so you write **1** in the answer space.

2. **Subtract the numbers in the second column.**

   Here, $5 - 5 = 0$, so you write **0** in the answer space.

3. Continue subtracting the columns, moving from right to left.

In the next column, you can't take 6 from 4 so you need to borrow from the 9. The 4 becomes 14, and the subtraction can then be completed — $14 - 6 = 8$, so you can write **8** in the answer space.

Next, $8 - 3 = 5$, so write **5** in the answer space.

You can't take 1 from 0 so you need to borrow from the 6. The 0 becomes 10, the subtraction can then be completed — $10 - 1 = 9$, so write **9** in the answer space.

In the next column, $5 - 4 = 1$, so write **1** in the answer space.

Finally, $1 - 0 = 1$, so write **1** in the answer space.

So $1{,}609{,}452 - 413{,}651 = 1{,}195{,}801$.

Multiplying

Multiplication is often described as a sort of shorthand for repeated addition. For example:

4×3 means add 4 to itself 3 times: $4 + 4 + 4 = 12$

9×6 means add 9 to itself 6 times: $9 + 9 + 9 + 9 + 9 + 9 = 54$

100×2 means add 100 to itself 2 times: $100 + 100 = 200$

Although multiplication isn't as warm and fuzzy as addition, it's a great timesaver. For example, suppose you play in a junior cricket team, and you've just won a game against the toughest team in the league. As a reward, your coach promised to buy three pies for each of the nine players on the team. To find out how many pies your coach needs, you can add 3 together 9 times. Or you can save time by multiplying 3 times 9, which gives you 27. Therefore, you need 27 pies (plus a whole lot of tomato sauce).

When you multiply two numbers, the two numbers that you're multiplying are called *factors*, and the result is the product.

In multiplication, the first number is also called the *multiplicand* and the second number is the *multiplier*. But almost nobody ever remembers — or uses — these words.

Signs of the times

When you're first introduced to multiplication, you use the times sign ($\times$). As you move onward and upward on your math journey, you need to be aware of the conventions I discuss in the following sections.

The symbol $\cdot$ is sometimes used to replace the symbol $\times$. For example,

$4 \cdot 2 = 8$	means	$4 \times 2 = 8$
$6 \cdot 7 = 42$	means	$6 \times 7 = 42$
$53 \cdot 11 = 583$	means	$53 \times 11 = 583$

In Part I of this book, I stick to the tried-and-true symbol $\times$ for multiplication. Just be aware that the symbol $\cdot$ exists so that you won't be stumped if your teacher or textbook uses it.

In maths beyond arithmetic, using parentheses without another operator stands for multiplication. The parentheses can enclose the first number, the second number, or both numbers. For example,

$3(5) = 15$	means	$3 \times 5 = 15$
$(8)7 = 56$	means	$8 \times 7 = 56$
$(9)(10) = 90$	means	$9 \times 10 = 90$

This switch makes sense when you stop to consider that the letter x, which is often used in algebra, looks a lot like the multiplication sign $\times$. So in this book, when I start using x in Part II, I also stop using $\times$ and begin using parentheses without another sign to indicate multiplication.

Memorising the multiplication table

You may consider yourself among the multiplicationally challenged. That is, you consider being called upon to remember 9×7 a tad less appealing than being dropped from an airplane while clutching a parachute purchased from the boot of some guy's car. If so, this section is for you.

Looking at the old multiplication table

One glance at the old multiplication table, Table 2-1, reveals the problem. If you saw the movie *Amadeus*, you may recall that Mozart was criticised for writing music that had 'too many notes'. Well, in my humble opinion, the multiplication table has too many numbers.

Table 2-1			The Monstrous Standard Multiplication Table							
	0	**1**	**2**	**3**	**4**	**5**	**6**	**7**	**8**	**9**
0	0	0	0	0	0	0	0	0	0	0
1	0	1	2	3	4	5	6	7	8	9
2	0	2	4	6	8	10	12	14	16	18
3	0	3	6	9	12	15	18	21	24	27
4	0	4	8	12	16	20	24	28	32	36
5	0	5	10	15	20	25	30	35	40	45
6	0	6	12	18	24	30	36	42	48	54
7	0	7	14	21	28	35	42	49	56	63
8	0	8	16	24	32	40	48	56	64	72
9	0	9	18	27	36	45	54	63	72	81

I don't like the multiplication table any more than you do. Just looking at it makes my eyes glaze over. With 100 numbers to memorise, no wonder so many people just give up and carry a calculator.

Introducing the short multiplication table

If the multiplication table from Table 2-1 were smaller and a little more manageable, I'd like it a lot more. So here's my short multiplication table, in Table 2-2.

Table 2-2		The Short Multiplication Table					
	3	*4*	*5*	*6*	*7*	*8*	*9*
3	9	12	15	18	21	24	27
4		16	20	24	28	32	36
5			25	30	35	40	45
6				36	42	48	54
7					49	56	63
8						64	72
9							81

As you can see, I've gotten rid of a bunch of numbers. In fact, I've reduced the table from 100 numbers to 28. I've also shaded 11 of the numbers I've kept.

Is just slashing and burning the sacred multiplication table wise? Is it even legal? Well, of course it is! After all, the table is just a tool, like a hammer. If a hammer's too heavy to pick up, you need to buy a lighter one. Similarly, if the multiplication table is too big to work with, you need a smaller one. Besides, I've removed only the numbers you don't need. For example, the condensed table doesn't include rows or columns for 0, 1, or 2. Here's why:

- Any number multiplied by 0 is 0 (people call this trait the *zero property of multiplication*).

- Any number multiplied by 1 is that number itself (which is why mathematicians call 1 the *multiplicative identity* — because when you multiply any number by 1, the answer is identical to the number you started with).

- Multiplying by 2 is fairly easy; if you can count by 2s — 2, 4, 6, 8, 10 and so forth — you can multiply by 2.

The rest of the numbers I've gotten rid of are redundant. (And not just redundant, but also repeated, extraneous and unnecessary!) For example, any way you slice it, 3×5 and 5×3 are both 15 (you can switch the order of the factors because multiplication is *commutative*). In my condensed table, I've simply removed the clutter.

So what's left? Just the numbers you need. These numbers include a grey row and a grey diagonal. The grey row is the 5 times table, which you probably know pretty well. (In fact, the 5s may evoke an early-childhood memory of running to find a hiding place on a warm spring day while one of your friends counted in a loud voice: 5, 10, 15, 20 ...)

The numbers on the grey diagonal are the square numbers — when you multiply any number by itself, the result is a square number. You probably know these numbers better than you think.

Getting to know the short multiplication table

In about an hour, you can make huge strides in memorising the multiplication table. To start, make a set of flash cards that give a multiplication problem on the front and the answer on the back. They may look like Figure 2-1.

Figure 2-1:
Both sides
of a flash
card, with
7 × 6 on the
front and 42
on the back.

Remember, you need to make only 28 flash cards — one for every example in Table 2-2. Split these 28 into two piles — a 'grey' pile with 11 cards and a 'white' pile with 17. (You don't have to colour the cards grey and white; just keep track of which pile is which, according to the shading in Table 2-2.) Then begin:

1. **5 minutes:** Work with the grey pile, going through it one card at a time. If you get the answer right, put that card on the bottom of the pile. If you get it wrong, put it in the middle so you get another chance at it more quickly.

2. **10 minutes:** Switch to the white pile and work with it in the same way.

3. **15 minutes:** Repeat Steps 1 and 2.

Now take a break. Really — the break is important to rest your brain. Come back later in the day and do the same thing.

When you're done with this exercise, you should find going through all 28 cards with almost no mistakes to be fairly easy. At this point, feel free to make cards for the rest of the standard times table — you know, the cards with all the 0, 1 and 2 times tables on them and the redundant problems — mix all 100 cards together, and amaze your family and friends.

To the nines: A slick trick

Here's a trick to help you remember the 9 times table. To multiply any one-digit number by 9:

1. **Subtract 1 from the number being multiplied by 9 and jot down the answer.**

   For example, suppose you want to multiply 7×9. Here, $7 - 1 = 6$.

2. **Jot down a second number so that, together, the two numbers you wrote add up to 9. You've just written the answer you were looking for.**

3. **Adding, you get $6 + 3 = 9$. So $7 \times 9 = 63$.**

As another example, suppose you want to multiply 8×9:

$$8 - 1 = 7$$

$$7 + 2 = 9$$

So $8 \times 9 = 72$.

This trick works for every one-digit number except 0 (but you already know that $0 \times 9 = 0$).

Double digits: Multiplying larger numbers

The main reason to know the multiplication table is so you can more easily multiply larger numbers. For example, suppose you want to multiply 53×7. Start by stacking these numbers on top of one another with a line underneath, and then multiply 3 by 7. Because $3 \times 7 = 21$, write down the 1 and carry the 2:

$$\begin{array}{r} 2 \\ 53 \\ \times 7 \\ \hline 1 \end{array}$$

Next, multiply 7 by 5. This time, $5 \times 7 = 35$. But you also need to add the 2 that you carried over, which makes the result 37. Because 5 and 7 are the last numbers to multiply, you don't have to carry, so write down the 37 — you find that $53 \times 7 = 371$:

$$\begin{array}{r} 2 \\ 53 \\ \times 7 \\ \hline 371 \end{array}$$

When multiplying larger numbers, the idea is similar. For example, suppose you want to multiply 53 by 47. (The first few steps — multiplying by the 7 in 47 — are the same, so I pick up with the next step.) Now you're ready to multiply by the 4 in 47. But remember that this 4 is in the tens column, so it really means 40. So to begin, put a 0 directly under the 1 in 371:

$$\begin{array}{r} 53 \\ \times 47 \\ \hline 371 \\ 20 \end{array}$$

This 0 acts as a placeholder so that this row is arranged properly.

When multiplying by larger numbers with two digits or more, use one placeholding zero when multiplying by the tens digit, two placeholding zeros when multiplying the hundreds digit, three zeros when multiplying by the thousands digit, and so forth.

Now you multiply 3×4 to get 12, so write down the 2 and carry the 1:

$$\begin{array}{r} 1 \\ 53 \\ \times 47 \\ \hline 371 \\ 20 \end{array}$$

Continuing, multiply 5×4 to get 20, and then add the 1 that you carried over, giving a result of 21:

$$\begin{array}{r} 1 \\ 53 \\ \times 47 \\ \hline 371 \\ 2120 \end{array}$$

To finish, add the two products (the multiplication results):

$$\begin{array}{r} 53 \\ \times 47 \\ \hline 371 \\ + \ 2120 \\ \hline 2,491 \end{array}$$

So $53 \times 47 = 2,491$.

Multiplication of larger numbers is performed in the same way as the multiplication of smaller numbers, with grid paper being even more useful.

You may find you move to working out larger multiplication problems on a calculator. As you do so, the communicative nature of multiplication is the most important thing to keep in mind — that is, that $4 \times 6 = 6 \times 4$. This rule is very useful when you reach the more complicated algebraic multiplication in later high school.

Doing Division Lickety-Split

The last of the Big Four operations is division. Division literally means splitting things up. For example, suppose you're on a picnic with two friends. You've brought along 12 cheese sticks as snacks, and want to split them fairly so that you and your two friends each get the same number (don't want to cause a fight, right?).

You each get four cheese sticks. This problem tells you that

$$12 \div 3 = 4$$

As with multiplication, division also has more than one sign: The division sign ($\div$) and the fraction slash (/) or fraction bar (—). So some other ways to write the same information are

$$12/3 = 4 \text{ and } \frac{12}{3} = 4$$

Whichever way you write it, the idea is the same: When you divide 12 cheese sticks equally among three people, each person gets 4 of them.

When you divide one number by another, the first number is called the *dividend*, the second is called the *divisor* and the result is the *quotient*. For example, in the division from the earlier example, the dividend is 12, the divisor is 3 and the quotient is 4.

Whatever happened to the division table?

Considering how much time teachers spend on the multiplication table, you may wonder why you've never seen a division table. For one thing, the multiplication table focuses on multiplying all the one-digit numbers by each other. This focus doesn't work too well for division because division usually involves at least one number that has more than one digit.

Besides, you can use the multiplication table for division, too, by reversing the way you normally use the table. For example, the multiplication table tells you that $6 \times 7 = 42$. You can reverse this equation to give you these two division problems:

$$42 \div 6 = 7$$
$$42 \div 7 = 6$$

Using the multiplication table in this way takes advantage of the fact that multiplication and division are *inverse operations*.

Making short work of long division

In the olden days, knowing how to divide large numbers — for example, $62{,}997 \div 843$ — was important. People used *long division*, an organised method for dividing a large number by another number. The process involved dividing, multiplying, subtracting and dropping numbers down.

But face it — one of the main reasons the pocket calculator was invented was to save 21st-century humans from ever having to do long division again.

Having said that, I need to add that your teacher and maths-crazy friends may not agree. Perhaps they just want to make sure you're not completely helpless if your calculator disappears somewhere into your backpack or the Bermuda Triangle. But if do you get stuck doing page after page of long division against your will, you have my deepest sympathy.

I will go this far, however: Understanding how to do long division with some not-too-horrible numbers is a good idea. In this section, I give you a good start with long division, telling you how to do a division problem that has a one-digit divisor.

Recall that the *divisor* in a division problem is the number that you're dividing by. When you're doing long division, the size of the divisor is your main concern: Small divisors are easy to work with, and large ones are a royal pain.

Working through an example

Suppose you want to find $860 \div 5$. Start off by writing the problem like this:

$$5\overline{)860}$$

Unlike the other main four operations, long division moves from left to right. In this case, you start with the number in the hundreds column (8). To begin, ask how many times 5 goes into 8 — that is, what's $8 \div 5$? The answer is 1 (with a little bit left over), so write 1 directly above the 8. Now multiply 1×5 to get 5, place the answer directly below the 8, and draw a line beneath it:

$$
\begin{array}{r}
1 \\
5\overline{)860} \\
\underline{5}
\end{array}
$$

Subtract $8 - 5$ to get 3. (**Note:** After you subtract, the result should always be smaller than the divisor. If not, you need to write a higher number above the division symbol.) Then bring down the 6 to make the new number 36:

$$
\begin{array}{r}
1 \\
5\overline{)860} \\
\underline{-5} \\
36
\end{array}
$$

These steps are one complete cycle — to complete the problem, you just need to repeat them. Now ask how many times 5 goes into 36 — that is, what's $36 \div 5$? The answer is 7 (with a little left over). Write 7 just above the 6, and then multiply 7×5 to get 35; write the answer under 36:

$$
\begin{array}{r}
17 \\
5\overline{)860} \\
\underline{-5} \\
36 \\
\underline{-35}
\end{array}
$$

Now subtract to get $36 - 35 = 1$; bring down the 0 next to the 1 to make the new number 10:

$$
\begin{array}{r}
172 \\
5\overline{)860} \\
-5 \\
\hline
36 \\
-35 \\
\hline
10
\end{array}
$$

Another cycle is complete, so begin the next cycle by asking how many times 5 goes into 10 — that is, $10 \div 5$. The answer this time is 2. Write down the 2 in the answer above the 0. Multiply to get $2 \times 5 = 10$, and write this answer below the 10:

$$
\begin{array}{r}
172 \\
5\overline{)860} \\
-5 \\
\hline
36 \\
-35 \\
\hline
10 \\
-10
\end{array}
$$

Now subtract $10 - 10 = 0$. Because you have no more numbers to bring down, you're finished, and here's the answer (that is, the quotient):

$$
\begin{array}{r}
172 \\
5\overline{)860} \\
-5 \\
\hline
36 \\
-35 \\
\hline
10 \\
-10 \\
\hline
0
\end{array}
$$

So $860 \div 5 = 172$.

This problem divides evenly, but many don't. If the number does not divide equally it can be written as a remainder or as a fraction or decimal.

Division of larger numbers is performed in the same way as the division of smaller numbers and, again, using grid paper can help you keep track of columns.

As you move to completing larger division problems on a calculator, still keep the long division process in mind. It comes in handy when you reach some of the more complicated algebraic division problems in later high school.

Chapter 3

Ups and Downs: Positive and Negative Numbers

In This Chapter

▶ Reading the signs of positive and negative numbers

▶ Adding, subtracting, multiplying and dividing signed numbers

▶ Understanding how zero affects signed numbers

*N*umbers have many characteristics: They can be big, little, even, odd, whole, fraction, positive, negative, and sometimes cold and indifferent. (I'm kidding about that last one.) 'In this chapter, I concentrate on mainly the positive and negative characteristics of numbers and how a number's sign reacts to different manipulations.

This chapter tells you how to add, subtract, multiply, and divide signed numbers, no matter whether all the numbers are all the same sign or a combination of positive and negative.

Showing Some Signs

Early on, mathematicians realised that using plus and minus signs and making rules for their use would be a big advantage in their number world. They also realised that if they used the minus sign, they wouldn't need to create a bunch of completely new symbols for negative numbers. After all, positive and negative numbers are related to one another, and inserting a minus sign in front of a number works well. Negative numbers have positive counterparts and vice versa.

Numbers that are opposite in sign but the same otherwise are *additive inverses*. Two numbers are additive inverses of one another if their sum is 0 — in other words, $a + (-a) = 0$. Additive inverses are always the same distance from 0 (in opposite directions) on the number line. For example, the additive inverse of -6 is $+6$; the additive inverse of $\frac{1}{5}$ is $-\frac{1}{5}$.

Picking out positive numbers

Positive numbers are greater than 0. They're on the opposite side of 0 from the negative numbers. If you were to arrange a tug-of-war between positive and negative numbers, the positive numbers would line up on the right side of 0, as shown in Figure 3-1.

Figure 3-1:
Positive
numbers
getting
larger to the
right.

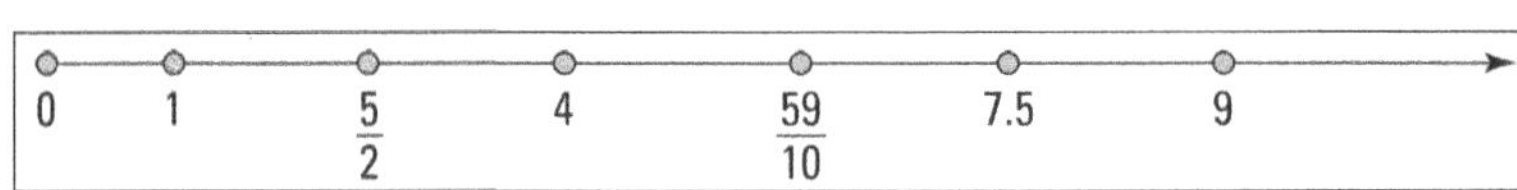

Positive numbers get bigger and bigger the farther they are from 0: 100°C, the boiling temperature of water, is hotter than 32°C, the temperature of a warm summer day, because 100 is farther away from 0 than 32 is. Both 100 and 32 are positive numbers, but one may seem 'more positive' than the other. Check out the difference between tepid water and boiling water to see how much more positive a number can be!

Making the most of negative numbers

The concept of a number less than 0 can be difficult to grasp. Sure, you can say 'less than 0', and even write a book with that title, but what does it really mean? Think of entering the ground floor of a large government building. You go to the elevator and have to choose between going up to the first, second, third or fourth floors, or going down to the first, second, third, fourth or fifth subbasement (down where all the secret stuff is). The farther you are from the ground floor, the farther the number of that floor is from 0. The second subbasement could be called floor –2, but that may not be a good number for a floor.

Negative numbers are smaller than 0. On a line with 0 in the middle, negative numbers line up on the left, as shown in Figure 3-2.

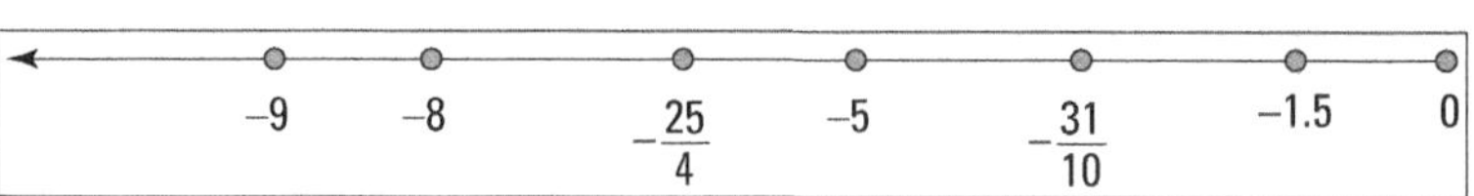

Figure 3-2: Negative numbers getting smaller to the left.

Negative numbers get smaller and smaller the farther they are from 0. This situation can get confusing because you may think that −400 is *bigger* than −12. But just think of −400°C and −12°C. Neither is anything pleasant to think about, but −400°C is definitely less pleasant — colder, lower, smaller.

When comparing negative numbers, the number closer to 0 is the bigger or greater number.

Comparing positives and negatives

Although my mum always told me not to compare myself to other people, comparing numbers to other numbers is often useful. And, when you compare numbers, the greater-than sign (>) and less-than sign (<) come in handy, which is why I use them in Table 3-1, where I put some positive- and negative-signed numbers in perspective.

Two other signs related to the greater-than and less-than signs are the greater-than-or-equal-to sign (≥) and the less-than-or-equal-to sign (≤).

Table 3-1	Comparing Positive and Negative Numbers
Comparison	*What It Means*
6 > 2	6 is greater than 2; 6 is farther from 0 than 2 is.
10 > 0	10 is greater than 0; 10 is positive and is bigger than 0.
−5 > −8	−5 is greater than −8; −5 is closer to 0 than −8 is.
−300 > −400	−300 is greater than −400; −300 is closer to 0 than −400 is.
0 > −6	Zero is greater than −6; −6 is negative and is smaller than 0.
7 > −80	7 is greater than −80. **Remember**: Positive numbers are always bigger than negative numbers.

So, putting the numbers 6, –2, –18, 3, 16 and –11 in order from smallest to biggest gives you: –18, –11, –2, 3, 6 and 16, which are shown as dots on a number line in Figure 3-3.

Figure 3-3:
Positive and negative numbers on a number line.

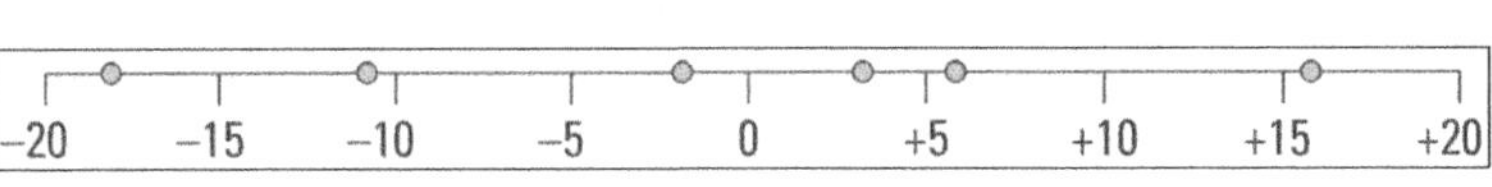

Zeroing in on zero

But what about 0? I keep comparing numbers to see how far they are from 0. Is 0 positive or negative? The answer is that it's neither. Zero has the unique distinction of being neither positive nor negative. Zero separates the positive numbers from the negative ones — what a job!

Operating with Signed Numbers

If you're on an elevator in a building with four floors above the ground floor and five floors below ground level, you can have a grand time riding the elevator all day, pushing buttons and actually 'operating' with signed numbers. If you want to go up five floors from the third subbasement, you end up on the second floor above ground level.

You're probably too young to remember this, but people actually used to get paid to ride elevators and push buttons all day. I wonder if these people had to understand algebra first ...

Adding like to like: Same-signed numbers

When your first-grade teacher taught you that $1 + 1 = 2$, she probably didn't tell you that this was just one part of the whole big addition story. She didn't mention that adding one positive number to another positive number is really a special case. If she *had* told you this big-story stuff — that you can add positive and negative numbers together or add any combination of positive and negative numbers together — you might have packed up your little school bag and lunchbox and left the room right then and there.

Adding positive numbers to positive numbers is just a small part of the whole addition story, but it was enough to get you started at that time. This section gives you the big story — all the information you need to add numbers of any sign. The first thing to consider in adding signed numbers is to start with the easiest situation — when the numbers have the same sign. Look at what happens:

- You have five CDs and your friend gives you four new CDs:

    $$(+5) + (+4) = +9$$

- You now have nine CDs.

- You owed Jon \$15 and had to borrow \$7 more from him:

    $$(-15) + (-7) = -22$$

- Now you're \$22 in debt.

You can use a nice 'S' rule for addition of positives to positives and negatives to negatives. See if you can say it quickly three times in a row: *When the signs are the same, you find the sum, and the sign of the sum is the same as the signs.* This rule holds when a and b represent any two real numbers:

$$(+a) + (+b) = +(a + b)$$
$$(-a) + (-b) = -(a + b)$$

I wish I had something as alliterative for all the rules, but this is maths, not poetry!

Say you're adding -3 and -2. The signs are the same; so you find the sum of 3 and 2, which is 5. The sign of this sum is the same as the signs of -3 and -2, so the *sum* is also a negative.

Here are some examples of finding the sums of same-signed numbers:

- **$(+4) + (+12) = +16$:** The signs are all positive.

- **$(-15) + (-85) = -100$:** The sign of the sum is the same as the signs.

- **$(+7) + (+3) + (+9) = +19$:** Because all the numbers are positive, add them and make the sum positive, too.

- **$(-2) + (-4) + (-5) + (-6) = -17$:** This time all the numbers are negative, so add them and give the sum a minus sign.

Adding same-signed numbers is a snap! (A little more alliteration for you.)

Adding different signs

Can a relationship between a Leo and a Gemini ever add up to anything? I don't know the answer to that question, but I do know that numbers with different signs add up very nicely. You just have to know how to do the computation and, in this section, I tell you.

When the signs of two numbers are different, forget the signs for a while and find the *difference* between the numbers.

Look what happens when you add numbers with different signs:

- You had $20 in your wallet and spent $15 for your theatre ticket:

    $(+20) + (-15) = +5$

- After settling up, you have $5 left.

- Your dad has $20, but he's put in $32 worth of petrol into his car:

    $(+20) + (-32) = -12$

- He'll have to borrow $12 to pay for the petrol.

Here are some examples of finding the sums of numbers with different signs:

- $(+5) + (-8) = -3$: The difference between 5 and 8 is 3. Eight is farther from 0 than 5 is, and 8 is negative, so the answer is -3.

- $(-5) + (+9) = +4$: This time 9 is positive. It's still farther from 0 than 5 is. The answer this time is $+4$.

- $(-5) + (+2) + (+11) + (-7) = +1$: If you take these in order from left to right (although you can add in any order you like), you add the first two together to get -3. Add -3 to the next number to get $+8$. Then add $+8$ to the last number to get $+1$.

Subtracting signed numbers

Subtracting signed numbers is really easy to do: You *don't*! Instead of inventing a new set of rules for subtracting signed numbers, mathematicians determined that it's easier to change the subtraction problems to addition problems and use the rules I explain in the previous section. Think of it as an original form of recycling.

Consider the method for subtracting signed numbers for a moment. Just change the subtraction problem into an addition problem? It doesn't make much sense, does it? Everybody knows that you can't just change an arithmetic operation and expect to get the same or right answer. You found out a long time ago that $10 - 4$ isn't the same as $10 + 4$. You can't just change the operation and expect it to come out correctly.

So, to make this work, you really change *two* things. (It almost seems to fly in the face of *two wrongs don't make a right*, doesn't it?)

When subtracting signed numbers, change the minus sign to a plus sign and change the number that the minus sign was in front of to its opposite. Then just add the numbers using the rules for adding signed numbers. For example:

- $(+a) - (+b) = (+a) + (-b)$
- $(+a) - (-b) = (+a) + (+b)$
- $(-a) - (+b) = (-a) + (-b)$
- $(-a) - (-b) = (-a) + (+b)$

The following examples put the process of subtracting signed numbers into real-life terms:

- The submarine was 50 metres below the surface when the skipper shouted, 'Dive!' It went down another 30 metres:

    $$-50 - (+30) = -50 + (-30) = -80$$

    Change from subtraction to addition. Change the 30 to its opposite, -30. Then use the addition rule. The submarine is now 80 metres below the surface.

- Some kids are pretending that they're on a reality-TV program and clinging to some footholds on a climbing wall. A team challenges the position of the opposing team's player. 'You were supposed to go down 3 metres, then up 8 metres, then down 4 metres. You shouldn't be 1 metre higher than you started!' The referee decides to check by having the player go backward — do the opposite moves. Making the player do the opposite, or subtracting the moves:

    $$-(-3) - (+8) - (-4) = +(+3) + (-8) + (+4) = -5 + (+4) = -1$$

    The player ended up 1 metre lower than where he started, so he had moved correctly in the first place.

Here are some examples of subtracting signed numbers:

- ✔ **–15 – 5 = –15 + (–5) = –20:** The subtraction becomes addition, and the +5 becomes negative. Then, because you're adding two signed numbers with the same sign, you find the sum and attach their common negative sign.

- ✔ **–4 – (–7) = –4 + (+7) = 3:** The subtraction becomes addition, and the –7 becomes positive. When adding numbers with opposite signs, you find their difference. The 3 is positive because the +7 is farther from 0.

- ✔ **9 – (–7) = 9 + (+7) = 16:** The subtraction becomes addition, and the –7 becomes positive. When adding numbers with the same sign, you find their sum. The two numbers are now both positive, so the answer is positive.

Multiplying and dividing signed numbers

Multiplication and division are really the easiest operations to do with signed numbers. As long as you can multiply and divide, the rules are not only simple, but the same for both operations.

Consider the stock market (something that gets considered a lot these days). The news reporter declares that the All Ordinaries went down 20 points twice in a row. You multiply two times –20 to get –40. So a positive times a negative is a negative.

Coming up with nothing

Consider adding two numbers with different signs where *no difference* exists between the absolute value of the numbers:

$$(+3) + (-3)$$
$$(-5) + (+5)$$

The difference between the numbers without their signs is 0. And because 0 is neither positive nor negative — it has no sign — that takes care of having to determine what the sign of the answer is by which number is farther from 0. Neither wins! So, in the following examples, 0 is the hero:

$$(-10) + (+10) = 0$$
$$(-a) + (+a) = 0$$
$$(+abc) + (-abc) = 0$$

In the last two examples, assume that a, b and c are the same throughout the expression.

How about dividing? You and three friends decide to buy another friend lunch. The luncher owes \$13.64. How much does each person chip in? Divide –13.64 by 4, and the –3.41 is what each of you contributes.

When multiplying and dividing two signed numbers, if the two signs are the same, the result is positive; when the two signs are different, the result is negative.

$$(+a) \times (+b) = +ab \qquad (+a) \div (+b) = + (a \div b)$$

$$(+a) \times (-b) = -ab \qquad (+a) \div (-b) = - (a \div b)$$

$$(-a) \times (+b) = -ab \qquad (-a) \div (+b) = - (a \div b)$$

$$(-a) \times (-b) = +ab \qquad (-a) \div (-b) = + (a \div b)$$

Notice in which cases the answer is positive and in which cases it's negative. You see that it doesn't matter whether the negative sign comes first or second when you have a positive and a negative. Also, notice that multiplication and division seem to be 'as usual' except for the positive and negative signs.

Here are some examples of multiplying and dividing signed numbers:

- $(-7) \times (+2) = -14$
- $(-8) \times (-11) = +88$
- $(+24) \div (-6) = -4$
- $(-50) \div (-25) = +2$

You can mix up these operations doing several multiplications or divisions or a mixture of each and use the following even-odd rule.

According to the even-odd rule, when multiplying and dividing a bunch of numbers, count the number of negatives to determine the final sign. An even number of negatives means the result is positive. An odd number of negatives means the result is negative.

Here are some examples of multiplying and dividing collections of signed numbers:

- $(+2) \times (-3) \times (+4) = -24$: This problem has just one negative sign. Because 1 is an odd number (and often the loneliest number), the answer is negative. The numerical parts (the 2, 3 and 4) get multiplied together and the negative is assigned as its sign.

✔ **(+2) × (–3) × (+4) × (–1) = +24:** Two negative signs mean a positive answer because 2 is an even number.

✔ **(–1)(–1)(–1)(–1) = 1:** An even number of negatives means you have a positive answer.

✔ **(–1)(–1)(–1)(–1)(–1)(–1)(–1)(–1)(–1)(–1)(–1)(–1)(–1)(–1)(–1) = –1:** An odd number of negative signs gives you a negative answer. All that negativity! And if there'd been just one more –1, the answer would've been positive ...

Working with Nothing: Zero and Signed Numbers

What role does 0 play in the signed-number show? What does 0 do to the signs of the answers? Well, when you're doing addition or subtraction, what 0 does depends on where it is in the problem. When you multiply or divide, 0 tends to just wipe out the numbers and leave you with nothing.

Here are some general guidelines about 0:

✔ **Adding zero:** $0 + a$ is just a. Zero doesn't change the value of a. (This is also true for $a + 0$.)

✔ **Subtracting zero:** $0 - a = -a$. Use the rule for subtracting signed numbers: Change the operation from subtraction to addition and change the sign of the second number, giving you $0 + (-a)$. But changing the order, $a - 0 = a$. It doesn't change the value of a to subtract 0 from it.

✔ **Multiplying by 0:** $a \times 0 = 0$. Twice nothing is nothing; three times nothing is nothing; multiply nothing and you get nothing. Likewise, $0 \times a = 0$.

✔ **Dividing 0 by a number:** $0 \div a = 0$. Take you and your friends: If none of you has anything, dividing that nothing into shares just means that each share has nothing.

You can't use 0 as a divisor. Numbers can't be divided by 0; not even 0 can be divided by 0. The answers just don't exist.

So, working with 0 isn't too tricky. You follow normal addition and subtraction rules, and just keep in mind that multiplying and dividing with 0 (0 being divided) leaves you with nothing — literally.

When using a scientific calculator, a graphic calculator or a CAS calculator, keep in mind that, generally, the negative sign has its own individual key. So take care when entering data onto your calculator, and make sure that you use the negative key to show a negative sign and not the subtraction or minus key.

Chapter 4

Parts of the Whole: Fractions, Decimals and Percentages

In This Chapter

▶ Completing the main four operations with fractions

▶ Working with decimals and the main operations

▶ Bringing in percentages

▶ Completing a refresher on converting between percentages, decimals and fractions

In this chapter, I first look at applying the main operations — addition, subtraction, multiplication and division — to fractions. I start by showing you how to multiply and divide fractions, which isn't much more difficult than multiplying whole numbers. Surprisingly, adding and subtracting fractions is a bit trickier. I show you a variety of methods, each with its own strengths and weaknesses, and I recommend how to choose which method will work best, depending on the problem you have to solve.

Later in the chapter, I move on to decimals and percentages, and here's some lovely news: Decimals are much easier to work with than fractions. Performing the main four operations on decimals is very close to performing them on whole numbers. The numerals 0 through 9 work just like they usually do so, as long as you get the decimal point in the right place, you're home free.

As well as covering working with fractions and decimals, I also show you how to convert fractions to decimals and decimals to fractions.

Multiplying and Dividing Fractions

One of the odd little ironies of life is that multiplying and dividing fractions is easier than adding or subtracting them — just two easy steps and you're done! For this reason, I show you how to multiply and divide fractions before I show you how to add or subtract them. In fact, you may find multiplying fractions easier than multiplying whole numbers because the numbers you're working with are usually small. More good news is that dividing fractions is nearly as easy as multiplying them. So I'm not even wishing you good luck — you don't need it!

Multiplying numerators and denominators straight across

Everything in life should be as simple as multiplying fractions. All you need for multiplying fractions is a pen or pencil, something to write on (preferably not your hand), and a basic knowledge of the multiplication table. (Refer to Chapter 2 for a multiplication refresher.)

Here's how to multiply two fractions:

1. **Multiply the numerators (the numbers on top) to get the numerator of the answer.**

2. **Multiply the denominators (the numbers on the bottom) to get the denominator of the answer.**

For example, here's how to multiply $\frac{2}{5} \times \frac{3}{7}$:

$$\frac{2}{5} \times \frac{3}{7} = \frac{2 \times 3}{5 \times 7} = \frac{6}{35}$$

Sometimes when you multiply fractions, you have an opportunity to reduce to lowest terms. As a rule, maths people are crazy about reduced fractions, and teachers sometimes take points off a right answer if you could've reduced it but didn't. Here's a multiplication problem that ends up with an answer that's not in its lowest terms:

$$\frac{4}{5} \times \frac{7}{8} = \frac{4 \times 7}{5 \times 8} = \frac{28}{40}$$

Because the numerator and the denominator are both even numbers, this fraction can be reduced. Start by dividing both numbers by 2:

$$\frac{28 \div 2}{40 \div 2} = \frac{14}{20}$$

Again, the numerator and the denominator are both even, so do it again:

$$\frac{14 \div 2}{20 \div 2} = \frac{7}{10}$$

This fraction is now fully reduced.

When multiplying fractions, you can often make your job easier by cancelling out equal factors in the numerator and denominator. Cancelling out equal factors makes the numbers that you're multiplying smaller and easier to work with, and it also saves you the trouble of reducing at the end. Here's how it works:

- When the numerator of one fraction and the denominator of the other are the same, change both of these numbers to 1s. (See the nearby sidebar for why this works.)
- When the numerator of one fraction and the denominator of the other are divisible by the same number, factor this number out of both. In other words, divide the numerator and denominator by that common factor.

For example, suppose you want to multiply the following two numbers:

$$\frac{5}{13} \times \frac{13}{20}$$

You can make this problem easier by cancelling out the number 13, as follows:

$$\frac{5}{1\ \cancel{13}} \times \frac{\cancel{13}\ 1}{20} = \frac{5 \times 1}{1 \times 20} = \frac{5}{20}$$

You can make it even easier by noticing that 20 and 5 are both divisible by 5, so you can also factor out the number 5 before multiplying:

$$\frac{1\ \cancel{5}}{1} \times \frac{1}{\cancel{20}\ 4} = \frac{1}{1} \times \frac{1}{4} = \frac{1}{4}$$

One is the easiest number

With fractions, the relationship between the numbers, not the actual numbers themselves, is most important. Understanding how to multiply and divide fractions can give you a deeper understanding of why you can increase or decrease the numbers within a fraction without changing the value of the whole fraction.

When you multiply or divide any number by 1, the answer is the same number. This rule also goes for fractions, so

$$\frac{3}{8} \times 1 = \frac{3}{8} \text{ and } \frac{3}{8} \div 1 = \frac{3}{8}$$

$$\frac{5}{13} \times 1 = \frac{5}{13} \text{ and } \frac{5}{13} \div 1 = \frac{5}{13}$$

$$\frac{67}{70} \times 1 = \frac{67}{70} \text{ and } \frac{67}{70} \div 1 = \frac{67}{70}$$

When a fraction has the same number in both the numerator and the denominator, its value is 1. In other words, the fractions $\frac{2}{2}$, $\frac{3}{3}$, and $\frac{4}{4}$ are all equal to 1. Look what happens when you multiply the fraction $\frac{3}{4}$ by $\frac{2}{2}$:

$$\frac{3}{4} \times \frac{2}{2} = \frac{3 \times 2}{4 \times 2} = \frac{6}{8}$$

The net effect is that you've increased the terms of the original fraction by 2. But all you've done is multiply the fraction by 1, so the value of the fraction hasn't changed. The fraction $\frac{6}{8}$ is equal to $\frac{3}{4}$.

Similarly, reducing the fraction $\frac{6}{9}$ by a factor of 3 is the same as dividing that fraction by $\frac{3}{3}$ (which is equal to 1):

$$\frac{6}{9} \div \frac{3}{3} = \frac{6 \div 3}{9 \div 3} = \frac{2}{3}$$

So $\frac{6}{9}$ is equal to $\frac{2}{3}$.

Multiplying mixed numbers

When multiplying mixed numbers, the process is always the same, and no quick tricks are available, sadly.

Say you're faced with the following example:

$$4\frac{4}{5} \times 1\frac{1}{3}$$

Here's how you complete the problem:

1. **Convert the mixed numbers into improper fractions before you do anything else.**

   Remember that an improper fraction is where the top number is larger than the bottom number.

$$4\frac{4}{5} \times 1\frac{1}{3} = \frac{24}{5} \times \frac{4}{3}$$

2. Look at the pairs of numbers diagonally across from each other and work out whether anything can be cancelled down.

That is, can anything be made smaller? In this example, 3 goes into 24, 8 times! (Phew — I don't want to multiply 24 by anything.) Cancel these numbers down, to be left with something that looks much friendlier.

$$= \frac{\cancel{24}^{8}}{5} \times \frac{4}{\cancel{3}^{1}}$$

$$= \frac{8}{5} \times \frac{4}{1}$$

3. Multiply across the top and then across the bottom.

$$= \frac{8}{5} \times \frac{4}{1}$$

$$= \frac{32}{5}$$

4. Simplify, if possible.

Remember that teachers are crazy about answers in simplest form. And in this example, you can simplify — 32 divided by 5 is 6 with 2 leftover.

$$4\frac{4}{5} \times 1\frac{1}{3} = \frac{32}{5}$$

$$= 6\frac{2}{5}$$

Doing a flip to divide fractions

Dividing fractions is just as easy as multiplying them. In fact, when you divide fractions, you really turn the problem into multiplication.

To divide one fraction by another, multiply the first fraction by the reciprocal of the second. (The *reciprocal* of a fraction is simply that fraction turned upside down.)

For example, here's how you turn fraction division into multiplication:

$$\frac{1}{3} \div \frac{4}{5} = \frac{1}{3} \times \frac{5}{4}$$

As you can see, I turn $\frac{4}{5}$ into its reciprocal — $\frac{5}{4}$ — and change the division sign to a multiplication sign. After that, just multiply the fractions as I describe in 'Multiplying numerators and denominators straight across':

$$\frac{1}{3} \times \frac{5}{4} = \frac{1 \times 5}{3 \times 4} = \frac{5}{12}$$

As with multiplication, in some cases, you may have to reduce your result at the end. But you can also make your job easier by cancelling out equal factors (see the preceding section).

Dividing mixed numbers

When dividing mixed numbers you need to change the mixed numbers to improper fractions *and* find the reciprocal of the second fraction *and* change the sign to a multiplication sign, *before* you can cancel anything out!

For example, say you're faced with the following.

$$3\frac{1}{3} \div 1\frac{1}{4}$$

Change to an improper fraction and find the reciprocal so you can multiply.

$$\frac{10}{3} \div \frac{5}{4}$$
$$= \frac{10}{3} \times \frac{4}{5}$$

Now you can see if you can cancel out so you're working with smaller numbers. In this case, you can simplify the 10 and the 5, by cancelling by 5.

$$= \frac{\cancel{10}^{2}}{3} \times \frac{4}{\cancel{5}^{1}}$$
$$= \frac{2}{3} \times \frac{4}{1}$$

Now simply multiply across the top and then the bottom. And, for the love of your teacher, simplify your answer.

$$3\frac{1}{3} \div 1\frac{1}{4} = \frac{2}{3} \times \frac{4}{1}$$
$$= \frac{8}{3}$$
$$= 2\frac{2}{3}$$

All Together Now: Adding Fractions

When you add fractions, one important item to notice is whether their denominators (the numbers on the bottom) are the same. If they're the same — woohoo! Adding fractions that have the same denominator is a walk in the park. But when fractions have different denominators, adding them becomes a tad more complex.

To make matters worse, many teachers make adding fractions even more difficult by requiring you to use a long and complicated method when, in many cases, a short and easy one will do.

In this section, I first show you how to add fractions with the same denominator. Then I show you a foolproof method for adding fractions when the denominators are different. It always works, and it's usually the simplest way to go. After that, I show you a quick method that you can use only for certain problems. Finally, I show you the longer, more complicated way to add fractions that usually gets taught.

Finding the sum of fractions with the same denominator

To add two fractions that have the same denominator (bottom number), add the numerators (top numbers) and leave the denominator unchanged.

For example, consider the following problem:

$$\frac{1}{5}+\frac{2}{5}=\frac{1+2}{5}=\frac{3}{5}$$

As you can see, to add these two fractions, you add the numerators $(1 + 2)$ and keep the denominator (5).

Why does this work? Keep in mind that you can think about fractions as pieces of cake. The denominator in this case tells you that the entire cake has been cut into five pieces. So when you add $\frac{1}{5}+\frac{2}{5}$, you're really adding one piece plus two pieces. The answer, of course, is three pieces — that is, $\frac{3}{5}$.

Even if you have to add more than two fractions, as long as the denominators are all the same, you just add the numerators and leave the denominator unchanged:

$$\frac{1}{17}+\frac{3}{17}+\frac{4}{17}+\frac{6}{17}=\frac{1+3+4+6}{17}=\frac{14}{17}$$

Sometimes when you add fractions with the same denominator, you have to reduce it to lowest terms. Take this problem, for example:

$$\frac{1}{4}+\frac{1}{4}=\frac{1+1}{4}=\frac{2}{4}$$

The numerator and the denominator are both even, so you know they can be reduced:

$$\frac{2}{4}=\frac{1}{2}$$

In other cases, the sum of two proper fractions is an improper fraction. You get a numerator that's larger than the denominator when the two fractions add up to more than 1, as in this case:

$$\frac{3}{7}+\frac{5}{7}=\frac{8}{7}$$

If you have more work to do with this fraction, leave it as an improper fraction so that it's easier to work with. But if this is your final answer, you may need to turn it into a mixed number:

$$\frac{8}{7}=8\div7=1\text{ remainder }1=1\frac{1}{7}$$

When two fractions have the same numerator, don't add them by adding the denominators and leaving the numerator unchanged.

Adding fractions with different denominators

When the fractions that you want to add have different denominators, adding them isn't quite as easy. At the same time, it doesn't have to be as hard as most teachers make it.

Now, I'm shimmying out onto a brittle limb here, but this needs to be said: Fractions can be added in a very simple way. It always works. It makes adding fractions only a little more difficult than multiplying them. And as you move up the maths food chain into algebra (the topic of Parts II and III of this book), it becomes the most useful method.

So why doesn't anybody talk about it? I think it's a clear case of tradition being stronger than common sense. The traditional way to add fractions is more difficult, more time-consuming, and more likely to cause an error. But generation after generation has been taught that it's the right way to add fractions. It's a vicious cycle.

But in this book, I'm breaking with tradition. I first show you the easy way to add fractions. Then I show you a quick trick that works in a few special cases. Finally, I show you the traditional way to add fractions.

Using the easy way

At some point in your life, I bet some teacher somewhere told you these golden words of wisdom: 'You can't add two fractions with different denominators.' Your teacher was wrong! Here's the way to do it:

1. **Cross-multiply the two fractions and add the results together to get the numerator of the answer.**

   Suppose you want to add the fractions $\frac{1}{3}$ and $\frac{2}{5}$. To get the numerator of the answer, cross-multiply. In other words, multiply the numerator of each fraction by the denominator of the other:

   $$\frac{1}{3} + \frac{2}{5}$$
   $$1 \times 5 = 5$$
   $$2 \times 3 = 6$$

   Add the results to get the numerator of the answer:

   $$5 + 6 = 11$$

2. **Multiply the two denominators to get the denominator of the answer.**

   To get the denominator, just multiply the denominators of the two fractions:

   $$3 \times 5 = 15$$

   The denominator of the answer is 15.

3. **Write your answer as a fraction.**

   $$\frac{1}{3} + \frac{2}{5} = \frac{11}{15}$$

As you discover in the earlier section 'Finding the sum of fractions with the same denominator', when you add fractions, you sometimes need to reduce the answer you get. Here's an example:

$$\frac{5}{8}+\frac{3}{10}=\frac{(5\times10)+(3\times8)}{8\times10}=\frac{50+24}{80}=\frac{74}{80}$$

Because the numerator and the denominator are both even numbers, you know that the fraction can be reduced. So try dividing both numbers by 2:

$$\frac{74\div2}{80\div2}=\frac{37}{40}$$

This fraction can't be reduced further, so $\frac{37}{40}$ is the final answer.

As you also discover in 'Finding the sum of fractions with the same denominator', sometimes when you add two proper fractions, your answer is an improper fraction:

$$\frac{4}{5}+\frac{3}{7}=\frac{(4\times7)+(3\times5)}{5\times7}=\frac{28+15}{35}=\frac{43}{35}$$

If you have more work to do with this fraction, leave it as an improper fraction so that it's easier to work with. But if this is your final answer, you may need to turn it into a mixed number.

$$\frac{43}{35}=43\div35=1\text{ remainder }8=1\frac{8}{35}$$

In some cases, you have to add more than one fraction. The method is similar, with one small tweak. For example, suppose you want to add $\frac{1}{2}+\frac{3}{5}+\frac{4}{7}$:

1. **Start by multiplying the *numerator* of the first fraction by the *denominators* of all the other fractions.**

$$\frac{1}{2}+\frac{3}{5}+\frac{4}{7}$$

$$(1\times5\times7)=35$$

2. **Do the same with the second fraction, and add this value to the first.**

$$\frac{1}{2}+\frac{3}{5}+\frac{4}{7}$$

$$35+(3\times2\times7)=35+42$$

3. Do the same with the remaining fraction(s).

$$\frac{1}{2}+\frac{3}{5}+\frac{4}{7}$$

$$35+42+(4\times2\times5)=35+42+40=117$$

When you're done, you have the numerator of the answer.

4. To get the denominator, just multiply all the denominators together:

$$\frac{1}{2}+\frac{3}{5}+\frac{4}{7}$$

$$=\frac{35+42+40}{2\times5\times7}=\frac{117}{70}$$

As usual, you may need to reduce or change an improper fraction to a mixed number. In this example, you just need to change to a mixed number:

$$\frac{117}{70}=117\div70=1\text{ remainder }47=1\frac{47}{70}$$

Trying a quick trick

I show you a way to add fractions with different denominators in the preceding section. It always works, and it's easy. So why do I want to show you another way? It feels like déjà vu.

In some cases, you can save yourself a lot of effort with a little bit of smart thinking. You can't always use this method, but you can use it when one denominator is a multiple of the other. Look at the following problem:

$$\frac{11}{12}+\frac{19}{24}$$

First, I solve it the way I show you in the preceding section:

$$\frac{11}{12}+\frac{19}{24}=\frac{(11\times24)+(19\times12)}{12\times24}=\frac{264+228}{288}=\frac{492}{288}$$

Those numbers are pretty big, and I'm still not done because the numerator is larger than the denominator. The answer is an improper fraction. Worse yet, the numerator and denominator are both even numbers, so the answer still needs to be reduced.

With certain fraction addition problems, I can give you a smarter way to work. The trick is to turn a problem with different denominators into a much easier problem with the same denominator.

Before you add two fractions with different denominators, check the denominators to see whether one is a multiple of the other. If it is, you can use the quick trick:

1. **Increase the terms of the fraction with the smaller denominator so that it has the larger denominator.**

   Look at the earlier problem in this new way:

   $$\frac{11}{12} + \frac{19}{24}$$

   As you can see, 12 divides into 24 without a remainder. In this case, you want to raise the terms of $\frac{11}{12}$ so that the denominator is 24:

   $$\frac{11}{12} = \frac{?}{24}$$

   To fill in the question mark, the trick is to divide 24 by 12 to find out how the denominators are related; then multiply the result by 11:

   $$? = (24 \div 12) \times 11 = 22$$

   $$\text{So } \frac{11}{12} = \frac{22}{24}$$

2. **Rewrite the problem, substituting this increased version of the fraction, and add as I show you earlier in 'Finding the sum of fractions with the same denominator'.**

   Now you can rewrite the problem this way:

   $$\frac{22}{24} + \frac{19}{24} = \frac{41}{24}$$

   As you can see, the numbers in this case are much smaller and easier to work with. The answer here is an improper fraction; changing it to a mixed number is easy:

   $$\frac{41}{24} = 41 \div 24 = 1 \text{ remainder } 17 = 1\frac{17}{24}$$

Relying on the traditional way

In the two preceding sections, I show you two ways to add fractions with different denominators. They both work great, depending on the circumstances. So why do I want to show you yet a third way? It feels like déjà vu all over again.

The truth is that I don't want to show you this way. But they're *forcing* me to. And you know who *they* are, don't you? The man — the system — the powers that be. The ones who want to keep you down in the mud, grovelling at their feet. Okay, so I'm exaggerating a little. But let me impress on you that you don't have to add fractions this way unless you really want to (or unless your teacher insists on it).

Here's the traditional way to add fractions with two different denominators:

1. **Find the least common multiple (LCM) of the two denominators.**

   Suppose you want to add the fractions $\frac{3}{4}+\frac{7}{10}$. First find the LCM of the two denominators, 4 and 10. Here's how to find the LCM using the multiplication table method:

   - **Multiples of 10:** 10, 20, 30, 40

   - **Multiples of 4:** 4, 8, 12, 16, 20

   So the LCM of 4 and 10 is 20.

2. **Increase the terms of each fraction so that the denominator of each equals the LCM.**

   Increase each fraction to higher terms so that the denominator of each is 20.

   $$\frac{3}{4}=\frac{3\times5}{4\times5}=\frac{15}{20} \text{ and } \frac{7}{10}=\frac{7\times2}{10\times2}=\frac{14}{20}$$

3. **Substitute these two new fractions for the original ones and add as I show you earlier in 'Finding the sum of fractions with the same denominator'.**

   At this point, you have two fractions that have the same denominator:

   $$\frac{15}{20}+\frac{14}{20}=\frac{29}{20}$$

   When the answer is an improper fraction, you still need to change it to a mixed number:

   $$\frac{29}{20}=29\div20=1 \text{ remainder } 9=1\frac{9}{20}$$

As another example, suppose you want to add the numbers $\frac{5}{6}+\frac{3}{10}+\frac{2}{15}$.

1. **Find the LCM of 6, 10, and 15.**

   This time, I use the prime factorisation method. Start by decomposing the three denominators to their prime factors:

   $$6=2\times3$$
   $$10=2\times5$$
   $$15=3\times5$$

These denominators have a total of three different prime factors — 2, 3, and 5. Each prime factor appears only once in any decomposition, so the LCM of 6, 10, and 15 is

$$2 \times 3 \times 5 = 30$$

2. **You need to increase the terms of all three fractions so that their denominators are 30:**

$$\frac{5}{6} = \frac{5 \times 5}{6 \times 5} = \frac{25}{30}$$

$$\frac{3}{10} = \frac{3 \times 3}{10 \times 3} = \frac{9}{30}$$

$$\frac{2}{15} = \frac{2 \times 2}{15 \times 2} = \frac{4}{30}$$

3. **Simply add the three new fractions:**

$$\frac{25}{30} = \frac{9}{30} + \frac{4}{30} = \frac{38}{30}$$

Again, you need to change this improper fraction to a mixed number:

$$\frac{38}{30} = 38 \div 30 = 1 \text{ remainder } 8 = 1\frac{8}{30}$$

Because both numbers are divisible by 2, you can reduce the fraction:

$$1\frac{8}{30} = 1\frac{4}{15}$$

Picking your trick: Choosing the best method

As I say earlier in this chapter, I think the traditional way to add fractions is more difficult than either the easy way or the quick trick. Your teacher may require you to use the traditional way, and after you get the hang of it, you'll get good at it. But given the choice, here's my recommendation:

✔ Use the easy way when the numerators and denominators are small (say, 15 or under).

✔ Use the quick trick with larger numerators and denominators when one denominator is a multiple of the other.

✔ Use the traditional way only when you can't use either of the other methods (or when you know the LCM just by looking at the denominators).

Adding mixed numbers with different denominators

You can use a sneaky short cut when adding mixed numbers. For example, say you need to work out the following.

$$3\frac{1}{2}+5\frac{2}{3}$$

First, take the whole numbers and add them together. In this case, $3 + 5 = 8$. Write this to the side of the question — highlight it or draw a box around it so that you don't forget about it!

$$= 8 + \frac{1}{2} + \frac{2}{3}$$

Now look at the fractions and add those using the method described in the previous section.

$$\frac{1}{2}+\frac{2}{3} = \frac{1\times3}{2\times3}+\frac{2\times2}{3\times2}$$
$$= \frac{3}{6}+\frac{4}{6}$$
$$= \frac{7}{6}$$
$$= 1\frac{1}{6}$$

Add back in the whole number that you already have to calculate your final answer. (And don't forget this step!)

$$3\frac{1}{2}+5\frac{2}{3} = 8 + 1\frac{1}{6}$$
$$= 9\frac{1}{6}$$

Unfortunately, this only works for adding mixed numbers; for all other calculations with mixed numbers you first have to change the mixed numbers to improper fractions.

Taking It Away: Subtracting Fractions

Subtracting fractions isn't really much different than adding them. As with addition, when the denominators are the same, subtraction is easy. And when the denominators are different, the methods I show you for adding fractions can be tweaked for subtracting them.

So to figure out how to subtract fractions, you can read the section 'All Together Now: Adding Fractions' and substitute a minus sign (–) for every plus sign (+). But it'd be just a little cheesy if I expected you to do that. So in this section, I show you four ways to subtract fractions that mirror what I discuss earlier in this chapter about adding them.

Subtracting fractions with the same denominator

As with addition, subtracting fractions with the same denominator is always easy. When the denominators are the same, you can just think of the fractions as pieces of cake.

To subtract one fraction from another when they both have the same denominator (bottom number), subtract the numerator (top number) of the second from the numerator of the first and keep the denominator the same. For example:

$$\frac{3}{5} - \frac{2}{5} = \frac{3-2}{5} = \frac{1}{5}$$

Sometimes, as when you add fractions, you have to reduce:

$$\frac{3}{10} - \frac{1}{10} = \frac{3-1}{10} = \frac{2}{10}$$

Because the numerator and denominator are both even, you can reduce this fraction by a factor of 2:

$$\frac{2}{10} = \frac{2 \div 2}{10 \div 2} = \frac{1}{5}$$

Unlike addition, when you subtract one proper fraction from another, you never get an improper fraction.

Subtracting fractions with different denominators

Just as with addition, you have a choice of methods when subtracting fractions. These three methods are similar to the methods I show you for adding fractions: The easy way, the quick trick and the traditional way.

The easy way always works, and I recommend this method for most of your fraction subtracting needs. The quick trick is a great timesaver, so use it when you can. And as for the traditional way — well, even if I don't like it, your teacher and other maths purists probably do.

Knowing the easy way

This way of subtracting fractions works in all cases, and it's easy. (In the next section, I show you a quick way to subtract fractions when one denominator is a multiple of the other.) Here's the easy way to subtract fractions that have different denominators:

1. **Cross-multiply the two fractions and subtract the second number from the first to get the numerator of the answer.**

   For example, suppose you want to subtract $\frac{6}{7} - \frac{2}{5}$. To get the numerator, cross-multiply the two fractions and then subtract the second number from the first number:

   $$\frac{6}{7} - \frac{2}{5}$$
   $$(6 \times 5) - (2 \times 7) = 30 - 14 = 16$$

   After you cross-multiply, be sure to subtract in the correct order. (The first number is the numerator of the first fraction times the denominator of the second.)

2. **Multiply the two denominators to get the denominator of the answer.**

   $$7 \times 5 = 32$$

3. **Putting the numerator over the denominator gives you your answer.**

   $$\frac{16}{35}$$

Here's another example to work with:

$$\frac{9}{10} - \frac{5}{6}$$

This time, I put all the steps together:

$$\frac{9}{10} - \frac{5}{6} = \frac{(9 \times 6) - (5 \times 10)}{10 \times 6}$$

With the problem set up like this, you just have to simplify the result:

$$= \frac{54 - 50}{60} = \frac{4}{60}$$

In this case, you can reduce the fraction:

$$\frac{4}{60} = \frac{1}{15}$$

Cutting it short with a quick trick

The easy way I show you in the preceding section works best when the numerators and denominators are small. When they're larger, you may be able to take a shortcut.

Before you subtract fractions with different denominators, check the denominators to see whether one is a multiple of the other. If it is, you can use the quick trick:

1. **Increase the terms of the fraction with the smaller denominator so that it has the larger denominator.**

   For example, suppose you want to find $\frac{17}{20} - \frac{31}{80}$. If you cross-multiply these fractions, your results are going to be much bigger than you want to work with. But fortunately, 80 is a multiple of 20, so you can use the quick way.

   First, increase the terms of $\frac{17}{20}$ so that the denominator is 80:

   $$\frac{17}{20} = \frac{?}{80}$$
   $$? = 80 \div 20 \times 17 = 68$$
   $$\text{So } \frac{17}{20} = \frac{68}{80}$$

2. **Rewrite the problem, substituting this increased version of the fraction, and subtract as I show you earlier in 'Subtracting fractions with the same denominator'.**

   Here's the problem as a subtraction of fractions with the same denominator, which is much easier to solve:

   $$\frac{68}{80} - \frac{31}{80} = \frac{37}{80}$$

   In this case, you don't have to reduce to lowest terms, although you may have to in other problems.

Keeping your teacher happy with the traditional way

As I describe earlier in this chapter in 'All Together Now: Adding Fractions', you want to use the traditional way only as a last resort. I recommend that you use it only when the numerator and denominator are too large to use the easy way and when you can't use the quick trick.

To use the traditional way to subtract fractions with two different denominators, follow these steps:

1. **Find the least common multiple (LCM) of the two denominators.**

   For example, suppose you want to subtract $\frac{7}{8} - \frac{11}{14}$. Here's how to find the LCM of 8 and 14:

   $$\text{Multiples of } 8 : 8, 16, 24, 40, 48, 56$$
   $$\text{Multiples of } 14 : 14, 28, 42, 56$$

   So the LCM of 8 and 14 is 56.

   $$\frac{7}{8} = \frac{7 \times 7}{8 \times 7} = \frac{49}{56}$$
   $$\frac{11}{14} = \frac{11 \times 4}{14 \times 4} = \frac{44}{56}$$

2. **Increase each fraction to higher terms so that the denominator of each equals the LCM.**

   The denominators of both now are 56:

   $$\frac{49}{56} - \frac{44}{56}$$

3. **Substitute these two new fractions for the original ones and subtract as I show you earlier in 'Subtracting fractions with the same denominator'.**

   $$\frac{49}{56} - \frac{44}{56} = \frac{5}{56}$$

   This time, you don't need to reduce because 5 is a prime number and 56 isn't divisible by 5. In some cases, however, you have to reduce the answer to lowest terms.

Subtracting mixed numbers with different denominators

Unfortunately, the trick I showed you for adding mixed numbers (refer to 'Adding mixed numbers with different denominators') does not help when subtracting mixed numbers. Say you're faced with the following.

$$5\frac{2}{7} - 3\frac{1}{2}$$

First, convert these mixed numbers to improper fractions.

$$= \frac{37}{7} - \frac{7}{2}$$

Then you can convert the denominators (the bottom numbers) to be the same.

$$= \frac{37^{\times 2}}{7^{\times 2}} - \frac{7^{\times 7}}{2^{\times 7}}$$

$$= \frac{74}{14} - \frac{49}{14}$$

Subtract the numerators (top numbers) and leave the bottom number as 14.

$$= \frac{74}{14} - \frac{49}{14}$$

$$= \frac{25}{14}$$

Now you need to convert back to a mixed number. Ask yourself: If I have 25 people, how many groups of 14 can I make? You can make one group, with 11 people left out.

$$5\frac{2}{7} - 3\frac{1}{2} = \frac{25}{14}$$

$$= 1\frac{11}{14}$$

Performing the Main Four Operations with Decimals

Everything you already know about adding, subtracting, multiplying, and dividing whole numbers (refer to Chapter 2) carries over when you work with decimals. In fact, in each case, there's really only one key difference: How to handle that pesky little decimal point. In this section, I show you how to perform the main four maths operations with decimals.

Adding decimals

Adding decimals is almost as easy as adding whole numbers. As long as you set up the problem correctly, you're in good shape. To add decimals, follow these steps:

1. **Arrange the numbers in a column and line up the decimal points vertically.**

2. **Add as usual, column by column, from right to left.**

3. **Place the decimal point in the answer in line with the other decimal points in the problem.**

For example, suppose you want to add the numbers 14.5 and 1.89. Line up the decimal points neatly, as follows:

```
  14.5
 +1.89
 ─────
```

Begin adding from the right-most column. Treat the blank space after 14.5 as a 0 — you can write this in as a trailing 0. Adding this column gives you $0 + 9 = 9$:

```
  14.50
 +1.89
 ─────
      9
```

Continuing to the left, $5 + 8 = 13$, so put down the 3 and carry the 1:

```
   1
  14.50
 +1.89
 ─────
     39
```

Complete the problem column by column and, at the end, put the decimal point directly below the others in the problem:

```
  14.50
 +1.89
 ─────
  16.39
```

When adding more than one decimal, the same rules apply. For example, suppose you want to add $15.1 + 0.005 + 800 + 1.2345$. The most important idea is lining up the decimal points correctly:

```
   15.1
    0.005
  800.0
 +1.2345
```

To avoid mistakes, be especially neat when adding a lot of decimals.

Because the number 800 isn't a decimal, I place a decimal point and a 0 at the end of it, to be clear about how to line it up. If you like, you can make sure all numbers have the same number of decimal places (in this case, four) by adding trailing zeros. When you properly set up the problem, the addition is no more difficult than in any other addition problem:

$$
\begin{array}{r}
15.1000 \\
0.0050 \\
800.0000 \\
+1.2345 \\
\hline
816.3395
\end{array}
$$

Subtracting decimals

Subtracting decimals uses the same trick as adding them (which I talk about in the preceding section). Here's how you subtract decimals:

1. **Arrange the numbers in a column and line up the decimal points.**

2. **Subtract as usual, column by column from right to left.**

3. **When you're done, place the decimal point in the answer in line with the other decimal points in the problem.**

For example, suppose you want to figure out $144.87 - 0.321$. First, line up the decimal points:

$$
\begin{array}{r}
144.870 \\
-0.321 \\
\hline
\end{array}
$$

In this case, I add a zero at the end of the first decimal. This placeholder reminds you that, in the right-most column, you need to borrow to get the answer to $0 - 1$:

$$
\begin{array}{r}
6 \\
144.8\cancel{7}\ 10 \\
-0.32\ \ 1 \\
\hline
4\ \ 9
\end{array}
$$

The rest of the problem is straightforward. Just finish the subtraction and drop the decimal point straight down:

$$
\begin{array}{r}
6 \\
144.8\cancel{7}\ 10 \\
-0.32\ \ 1 \\
\hline
144.54\ \ 9
\end{array}
$$

Therefore 144.87 – 0.321 = 144.549. As with addition, the decimal point in the answer goes directly below where it appears in the problem.

Multiplying decimals

Multiplying decimals is different from adding and subtracting them, in that you don't have to worry about lining up the decimal points (see the preceding sections). In fact, the only difference between multiplying whole numbers and decimals comes at the end.

Here's how to multiply decimals:

1. **Perform the multiplication as you do for whole numbers.**

2. **When you're done, count the number of digits to the right of the decimal point in each factor, and add the result.**

3. **Place the decimal point in your answer so that your answer has the same number of digits after the decimal point.**

This process sounds tricky, but multiplying decimals can actually be simpler than adding or subtracting them. Suppose, for instance, that you want to multiply 23.5 by 0.16. The first step is to pretend that you're multiplying numbers without decimal points (that is, 235 multiplied by 16):

```
   23.5
 ×0.16
 ─────
  1410
  2350
 ─────
  3760
```

This answer isn't complete, though, because you still need to find out where the decimal point goes in the answer. To do this, notice that 23.5 has one digit after the decimal point and that 0.16 has two digits after the decimal point. Because 1 + 2 = 3, place the decimal point in the answer so that it has three digits after the decimal point. (You can put your pencil at the 0 at the end of 3760 and move the decimal point three places to the left.)

```
   23.5   1 digit after the decimal point
 ×0.16   2 digits after the decimal point
  1410
  2350
 ─────
  3760   1 + 2 = 3 digits after the decimal point
```

Even though the last digit in the answer is a 0, you still need to count this as a digit when placing the decimal point. When the decimal point is in place, you can drop trailing zeros.

So the answer is 3.760, which is equal to 3.76.

Dividing decimals

Long division has never been a crowd pleaser. Dividing decimals is almost the same as dividing whole numbers, which is why a lot of people don't particularly like dividing decimals, either.

But at least you can take comfort in the fact that, when you know how to do long division (which I cover in Chapter 2), figuring out how to divide decimals is easy. The main difference comes at the beginning, before you start dividing.

Here's how to divide decimals:

1. **Turn the *divisor* (the number you're dividing by) into a whole number by moving the decimal point all the way to the right; at the same time, move the decimal point in the *dividend* (the number you're dividing) the same number of places to the right.**

   For example, suppose you want to divide 10.274 by 0.11. Write the problem as usual:

   $$0.11 \overline{)10.274}$$

   Turn 0.11 into a whole number by moving the decimal point in 0.11 two places to the right, giving you 11. At the same time, move the decimal point in 10.274 two places to the right, giving you 1,027.4:

   $$11 \overline{)1027.4}$$

2. **Place a decimal point in the *quotient* (the answer) directly above where the decimal point now appears in the dividend.**

   Here's what this step looks like:

   $$11. \overline{)1027.4}$$

3. **Divide as usual, being careful to line up the quotient properly so that the decimal point falls into place.**

   To start out, notice that 11 is too large to go into either 1 or 10. However, 11 does go into 102 (nine times). So write the first digit of the quotient just above the 2 and continue:

```
       9.
11.)1027.4
    99
    37
```

I paused after bringing down the next number, 7. This time, 11 goes
into 37 three times. The important point is to place the next digit in the
answer just above the 7:

```
      93.
11.)1027.4
    99
    37
    33
    44
```

I paused after bringing down the next number, 4. Now, 11 goes into 44
four times. Again, be careful to place the next digit in the quotient just
above the 4, and complete the division:

```
      93.4
11.)1027.4
    99
    37
    33
    44
    44
     0
```

So the answer is 93.4. As you can see, as long as you're careful when
placing the decimal point and the digits, the correct answer appears
with the decimal point in the right position.

Dealing with more zeros in the dividend

Sometimes you have to add one or more trailing zeros to the dividend.
Remember — you can add as many trailing zeros as you like to a decimal
without changing its value. For example, suppose you want to divide 67.8
by 0.333:

```
0.333)67.8
```

Follow these steps:

1. **Change 0.333 into a whole number by moving the decimal point three places to the right; at the same time, move the decimal point in 67.8 three places to the right:**

   $$333.\overline{)67800.}$$

   In this case, when you move the decimal point in 67.8, you run out of room, so you have to add a couple zeros to the dividend. This step is perfectly valid, and you need to do this whenever the divisor has more decimal places than the dividend.

2. **Place the decimal point in the quotient directly above where it appears in the dividend:**

   $$333.\overline{)67800.}$$

3. **Divide as usual, being careful to correctly line up the numbers in the quotient. This time, 333 doesn't go into 6 or 67, but it does go into 678 (two times). So place the first digit of the quotient directly above the 8:**

   $$
   \begin{array}{r}
   2 \\
   333.\overline{)67800.} \\
   \underline{666} \\
   120
   \end{array}
   $$

   I've jumped forward in the division to the place where I bring down the first 0. At this point, 333 doesn't go into 120, so you need to put a 0 above the first 0 in 67,800 and bring down the second 0. Now, 333 does go into 1,200, so place the next digit in the answer (3) over the second 0:

   $$
   \begin{array}{r}
   203 \\
   333.\overline{)67800.} \\
   \underline{666} \\
   1200 \\
   \underline{999} \\
   201
   \end{array}
   $$

This time, the division doesn't work out evenly. If this were a problem with whole numbers, you'd finish by writing down a remainder of 201. But decimals are a different story. The next section explains why, with decimals, the show must go on.

Completing decimal division

When you're dividing whole numbers, you can complete the problem simply by writing down the remainder. But remainders are *never* allowed in decimal division.

A common way to complete a problem in decimal division is to round off the answer. In most cases, you're instructed to round your answer to the nearest whole number or to one or two decimal places.

To complete a decimal division problem by rounding it off, you need to add at least one trailing zero to the dividend:

- ✔ To round a decimal to a whole number, add one trailing zero.
- ✔ To round a decimal to one decimal place, add two trailing zeros.
- ✔ To round a decimal to two decimal places, add three trailing zeros.

Here's what the problem from the preceding section looks like with a trailing zero attached:

$$
\begin{array}{r}
203\,. \\
333.\overline{)67800.} \\
\underline{666} \\
1200 \\
\underline{999} \\
2010
\end{array}
$$

Attaching a trailing zero doesn't change a decimal, but it does allow you to bring down one more number, changing 201 into 2,010. Now you can divide 333 into 2,010:

$$
\begin{array}{r}
203.6 \\
333.\overline{)67800.0} \\
\underline{666} \\
1200 \\
\underline{999} \\
2010 \\
\underline{1998} \\
12
\end{array}
$$

At this point, you can round the answer to the nearest whole number, 204.

Rounding off correctly

Rounding can be tricky, but basically you have one rule to remember: Look at the digit directly behind the decimal place that you need to round to and then round as follows:

- ✔ If the digit is between 0 and 4, the digit in the place to which you are rounding stays the same.
- ✔ If the digit is between 5 and 9, the digit in the place to which you are rounding goes up one.

Some examples may make this rule clearer. Say you need to round 2.35**6**78 to two decimal places. Look at the digit in the third place (bolded here). Because 6 lies between 5 and 9, the second digit goes up one. This means 2.35678 is rounded to 2.36.

Now look at rounding 2.3**1** to one decimal place. Look at the digit in the second place (again, bolded). Because 1 lies between 0 and 4, the first digit stays the same. So 2.31 is rounded to 2.3.

If a question asks you to round to two decimal places and the answer only has one decimal place, you can add a trailing zero to fill that second place. For example, if your answer is 1.4 and your teacher asks you to answer to two decimal places, your answer becomes 1.40.

Now look at a more difficult example. Say you need to round 2.899 to two decimal places. The digit in the third place lies between 5 and 9, so the second digit goes up one. In this case, if the second digit goes up one, it becomes a 10, affecting the digit in front of it. So rounding 2.899 to two decimal places makes it 2.90.

Checking your answers

When working with whole numbers or decimals (as with all fractions), it is always important to check your answer. You can do this by estimating an answer and then checking your answer against the estimation. Ask yourself: Is your answer reasonable?

For example, if you're dividing 101.75 by 24, what would you expect your answer to be?

In this case, 101.75 is close to 100 and 24 is close to 25 and $100 \div 25 = 4$, so you would expect your answer to be close to 4. If your answer is 24.6 you know you've made a mistake and you need to go back and check your work.

Never ever, ever hand your test in early — always use any spare time to check your answers and make sure that they are all reasonable and what you expected.

Converting to and from Percentages, Decimals and Fractions

To solve many percentage problems, you need to change the percentage to either a decimal or a fraction. Then you can apply what you know about solving decimal and fraction problems. For this reason, I show you how to convert to and from percentages.

Percentages and decimals are similar ways of expressing parts of a whole. This similarity makes converting percentages to decimals, and vice versa, mostly a matter of moving the decimal point. It's so simple you can probably do it in your sleep (but you should probably stay awake when you first read about the concept).

Percentages and fractions both express the same idea — parts of a whole — in different ways. So converting back and forth between percentages and fractions isn't quite as simple as just moving the decimal point back and forth. In this section, I cover the ways to convert to and from percentages, decimals and fractions, starting with percentages to decimals.

Going from percentages to decimals

To convert a percentage to a decimal, drop the per cent sign (%) and move the decimal point two places to the left. It's simple as you are actually dividing by 100. Remember that, in a whole number, the decimal point comes at the end. For example,

$$2.5\% = 0.025$$
$$4\% = 0.04$$
$$36\% = 0.36$$
$$111\% = 1.11$$

Changing decimals into percentages

To convert a decimal to a percentage, move the decimal point two places to the right as you are actually multiplying by 100, and add a per cent sign (%):

$$0.07 = 7\%$$
$$0.21 = 21\%$$
$$0.375 = 37.5\%$$

Switching from percentages to fractions

Converting percentages to fractions is fairly straightforward. Remember that the term 'per cent' means 'out of 100'. So changing percentages to fractions naturally involves the number 100.

To convert a percentage to a fraction, use the number in the percentage as your numerator (top number) and the number 100 as your denominator (bottom number):

$$39\% = \frac{39}{100} \qquad 86\% = \frac{86}{100} \qquad 217\% = \frac{217}{100}$$

As always with fractions, you may need to reduce to lowest terms or convert an improper fraction to a mixed number.

In the three examples, $\frac{39}{100}$ can't be reduced or converted to a mixed number. However, $\frac{86}{100}$ can be reduced because the numerator and denominator are both even numbers:

$$\frac{86}{100} = \frac{43}{50}$$

And $\frac{217}{100}$ can be converted to a mixed number because the numerator (217) is greater than the denominator (100):

$$\frac{217}{100} = 2\frac{17}{100}$$

Once in a while, you may start out with a percentage that's a decimal, such as 99.9%. The rule is still the same, but now you have a decimal in the numerator (top number), which most people don't like to see. To get rid of it, move the decimal point one place to the right in both the numerator and the denominator:

$$99.9\% = \frac{99.9}{100} = \frac{999}{1000}$$

Thus, 99.9% converts to the fraction $\frac{999}{1000}$.

Turning fractions into percentages

Converting a fraction to a percentage is really a two-step process. Here's how to do it:

1. **Convert the fraction to a decimal.**

   For example, suppose you want to convert the fraction $\frac{4}{5}$ to a percentage. To convert $\frac{4}{5}$ to a decimal, you can divide the numerator by the denominator:

   $$\frac{4}{5} = 0.8$$

2. **Convert this decimal to a percentage.**

   Convert 0.8 to a percentage by moving the decimal point two places to the right and adding a per cent sign (as I show you earlier in 'Changing decimals into percentages').

   $$0.8 = 80\%$$

Now suppose you want to convert the fraction $\frac{5}{8}$ to a percentage. Follow these steps:

1. **Convert $\frac{5}{8}$ to a decimal by dividing the numerator by the denominator:**

$$
\begin{array}{r}
0.625 \\
8\overline{)5.000} \\
\underline{48} \\
20 \\
\underline{16} \\
40 \\
\underline{40} \\
0
\end{array}
$$

   Therefore, $\dfrac{5}{8} = 0.625$.

2. **Convert 0.625 to a percentage by moving the decimal point two places to the right and adding a per cent sign (%):**

$$0.625 = 62.5\%$$

Most of these calculations can be easily performed on a scientific or CAS calculator. Having a good understanding of the process behind these answers, however, is still important so that you're ready for the complex algebra heading your way.

Chapter 5

Understanding Order of Operations

In This Chapter

▶ Organising your operations into bite-size pieces

▶ Ordering your problem from first to last

▶ Making sure your answers make sense

Algebra had its start as expressions that were all words. Everything was literally spelled out. As symbols and letters were added, algebraic manipulations became easier. But, as more symbols and notations were added, the rules that went along with the symbols also became a part of algebra. All this shorthand is wonderful, as long as you know the rules and follow the steps that go along with them. The order of operations is a biggie that you use frequently when working in algebra. It tells you what to do first, next and last in a problem, whether terms are in grouping symbols or raised to a power.

And, because you may not always remember the order of operations correctly, checking your work is very important. Making sure that the answer you get makes sense, and that it actually solves the problem, is an important step of working every problem.

This chapter walks you through the order of operations and checking your answers.

Ordering Operations

When does it matter in what *order* you do things? Or does it matter at all? Well, take a look at a couple of real-world situations:

▶ When you're cleaning your bedroom, it *doesn't* matter whether you make the bed or take your dirty clothes out to the laundry first. (I'm sure your parents would just love you to do either.)

✔ When you're getting dressed, it *does* matter whether you put on your shoes first or your socks first.

Sometimes the order matters, sometimes it doesn't. In algebra, the order depends on which mathematical operations are performed. If you're doing only addition or you're doing only multiplication, you can use any order you want. But as soon as you mix things up with addition and multiplication in the same expression, you have to pay close attention to the correct order. You can't just pick and choose what to do first, next and last according to what you feel like doing.

For example, look at the different ways this problem could be done, if no rules existed. Notice that all four operations are represented here.

$$19 - 4 \times 4 + 6 \div 2 =$$

One way to do the problem is to just go from left to right:

1. $19 - 4 = 15$
2. $15 \times 4 = 60$
3. $60 + 6 = 66$
4. $66 \div 2 = 33$

This gives you a final answer of 33. Unfortunately, this is incorrect.

Another approach is to group the 4×4 together in parentheses. Grouped terms tell you that you have to do the operation inside the grouping symbol first (for more on this, see 'Gathering Terms with Grouping Symbols', later in this chapter).

1. $19 - (4 \times 4) = 19 - 16 = 3$
2. $3 + 6 = 9$
3. $9 \div 2 = 4\frac{1}{2}$

This gives you a final answer of $4\frac{1}{2}$. This is incorrect also. The correct answer is actually 6. You can come back and check it once you have finished reading this chapter.

Using other groupings, I can make the answer come out to be different. I won't go into how these answers are obtained because they're all wrong anyway.

Mathematicians designed rules so that anyone reading a mathematical expression would do it the same way as everyone else and get the same *correct* answer. In the case of multiple signs and operations, working out the problems needs to be done in a specified *order*, from the first to the last. This is the *order of operations*.

According to the order of operations, work out the operations and signs in the following order:

1. **Powers and roots**

2. **Multiplication and division**

3. **Addition and subtraction**

If you have more than two operations of the same level, do them in order from left to right, following the order of operations. The following sections provide more detail on this.

Applying order of operations to the main four expressions

Generally speaking, the main four expressions come in the three types in Table 5-1.

Table 5-1	The Three Types of Main Four Expressions	
Expression	*Example*	*Rule*
Contains only addition and subtraction	$12 + 7 - 6 - 3 + 8$	Evaluate left to right.
Contains only multiplication and division	$18 \div 3 \times 7 \div 14$	Evaluate left to right.
Mixed-operator expression: Contains a combination of addition/subtraction and multiplication/division	$9 + 6 \div 3$	1. Evaluate multiplication and division left to right. 2. Evaluate addition and subtraction left to right.

In this section, I show you how to identify and evaluate all three types of expressions.

Expressions with only addition and subtraction

Some expressions contain only addition and subtraction. When this is the case, the rule for evaluating the expression is simple.

When an expression contains only addition and subtraction, evaluate it step by step from left to right. For example, suppose you want to evaluate this expression:

$$17 - 5 + 3 - 8$$

Because the only operations are addition and subtraction, you can evaluate from left to right, starting with $17 - 5$:

$$= 12 + 3 - 8$$

As you can see, the number 12 replaces $17 - 5$. Now the expression has three numbers instead of four. Next, evaluate $12 + 3$:

$$= 15 - 8$$

This step breaks down the expression to two numbers, which you can evaluate easily:

$$= 7$$

So $17 - 5 + 3 - 8 = 7$.

Expressions with only multiplication and division

Some expressions contain only multiplication and division. When this is the case, the rule for evaluating the expression is pretty straightforward.

When an expression contains only multiplication and division, evaluate it step by step from left to right. Suppose you want to evaluate this expression:

$$9 \times 2 \div 6 \div 3 \times 2$$

Again, the expression contains only multiplication and division, so you can move from left to right, starting with 9×2:

$$= 18 \div 6 \div 3 \times 2$$
$$= 3 \div 3 \times 2$$
$$= 1 \times 2$$
$$= 2$$

Notice that the expression shrinks one number at a time until all that's left is 2. So $9 \times 2 \div 6 \div 3 \times 2 = 2$.

Here's another quick example:

$$-2 \times 6 \div -4$$

Even though this expression has some negative numbers, the only operations it contains are multiplication and division. So you can evaluate it in two steps from left to right (remembering the rules for multiplying and dividing with negative numbers that I show you in Chapter 3):

$$= -2 \times 6 \div -4$$
$$= -12 \div -4$$
$$= 3$$

Thus, $= 2 \times 6 \div -4 = 3$.

Mixed-operator expressions

Often an expression contains

- ✔ At least one addition or subtraction operator.
- ✔ At least one multiplication or division operator.

I call these *mixed-operator expressions*. To evaluate them, you need some stronger medicine.

Evaluate mixed-operator expressions as follows:

1. **Evaluate the multiplication and division from left to right.**
2. **Evaluate the addition and subtraction from left to right.**

For example, suppose you want to evaluate the following expression:

$$5 + 3 \times 2 + 8 \div 4$$

As you can see, this expression contains addition, multiplication and division, so it's a mixed-operator expression. To evaluate it, start by underlining the multiplication and division in the expression:

$$5 + \underline{3 \times 2} + \underline{8 \div 4}$$

Now evaluate what you've underlined from left to right:

$$= 5 + 6 + \underline{8 \div 4}$$
$$= 5 + 6 + 2$$

At this point, you're left with an expression that contains only addition, so you can evaluate it from left to right:

$$= 11 + 2$$
$$= 13$$

Thus, $5 + 3 \times 2 + 8 \div 4 = 13$.

Using order of operations in expressions with exponents and roots

Here's what you need to know to evaluate expressions that have exponents and roots.

Evaluate exponents and roots from left to right *before* you begin evaluating the main four operations (adding, subtracting, multiplying and dividing).

The trick here is to turn the expression into a main four expression and then use what I show you earlier in 'Applying order of operations to Big Four expressions'. For example, suppose you want to evaluate the following:

$$3+5^2-6$$

First, evaluate the exponent:

$$3+25-6$$

At this point, the expression contains only addition and subtraction, so you can evaluate it from left to right in two steps:

$$= 28 - 6$$
$$= 22$$

So $3+5^2-6 = 22$.

Here's another example. Simplify the following expression using the order of operations:

$$6^2 - 5 \times 4 + 2\sqrt{16}$$

Perform the power and root first:

$$6^2 - 5 \times 4 + 2\sqrt{16} = 36 - 5 \times 4 + 2 \times 4$$

A multiplication symbol is introduced when the radical is removed — to show that the 2 multiplies the result. Two multiplications are performed to get 36 − 20 + 8. Now subtract and add: 16 + 8 = 24.

Grouping symbols not carrying their weight

When writing algebraic expressions, you use parentheses, brackets and braces to show what operations need to be performed first. When several different groupings are necessary, you use more than one type, usually nested one within the other, to make reading the expression easier. You can't do this with graphing calculators, though. The brackets and braces mean something entirely different in those instruments. In most graphing calculators:

✔ **Brackets** mean that the items inside are a part of a matrix, a rectangular arrangement of numbers.

✔ **Braces** mean that what's inside is part of a list of numbers.

These differences make for some awkward situations when you want to show several groupings within a single expression. Because you're limited to parentheses only, and they're all the same size, there's often confusion as to when a grouping starts and where it ends. You trade the convenience of the calculator for the inconvenience of the notation.

Gathering Terms with Grouping Symbols

In algebra problems, parentheses, brackets and braces are all used for grouping. Terms inside the grouping symbols have to be operated upon before they can be acted upon by anything outside the grouping symbol. All the grouping types have equal weight; none is more powerful or acts differently from the others.

If the problem contains grouped items, do what's inside a grouping symbol first, and then follow the order of operations. The grouping symbols are

✔ **Parentheses ():** Parentheses are the most commonly used symbols for grouping.

✔ **Brackets [ ] and braces { }:** Brackets and braces are also used frequently for grouping and have the same effect as parentheses. Using the different types of symbols helps when a problem uses more than one grouping. It's easier to tell where a group starts and ends.

✔ **Radical $\sqrt{}$:** This is an operation used for finding roots.

✔ **Fraction line (called the *vinculum*):** The fraction line also acts as a grouping symbol; everything above the line in the numerator is grouped together, and everything below the line in the denominator is grouped together.

Understanding order of precedence in expressions with parentheses

When it comes to evaluating expressions, here's what you need to know about parentheses.

To evaluate expressions that contain parentheses:

1. **Evaluate the contents of parentheses from the inside out.**

2. **Evaluate the rest of the expression.**

Main four expressions with parentheses

Similarly, suppose you want to evaluate $(1+15 \div 5)+(3-6)\times 5$. This expression contains two sets of parentheses, so evaluate these from left to right. Notice that the first set of parentheses contains a mixed-operator expression, so evaluate this in two steps, starting with the division:

$$= (1+3)+(3-6)\times 5$$
$$= 4+(3-6)\times 5$$

Now evaluate the contents of the second set of parentheses:

$$= 4+-3\times 5$$

Now you have a mixed-operator expression, so evaluate the multiplication (-3×5) first:

$$= 4 + -15$$

Finally, evaluate the addition:

$$= -11$$

So $(1+15 \div 5)+(3-6)\times 5 = -11$.

Expressions with exponents and parentheses

As another example, try this out:

$$1+(3-6^2 \div 9)\times 2^2$$

Start by working with *only* what's inside the parentheses. The first part to evaluate there is the exponent, 6^2:

$$= 1+(3-36 \div 9)\times 2^2$$

Continue working inside the parentheses by evaluating the division $36 \div 9$:

$$= 1 + (3 - 4) \times 2^2$$

Now you can get rid of the parentheses altogether:

$$= 1 - 1 \times 2^2$$

At this point, what's left is an expression with an exponent. This expression takes three steps, starting with the exponent:

$$= 1 - 1 \times 4$$
$$= 1 - 4$$
$$= -3$$

So $1 + (3 - 6^2 \div 9) \times 2^2 = -3$.

Expressions with parentheses raised to an exponent

Sometimes the entire contents of a set of parentheses are raised to an exponent. In this case, evaluate the contents of the parentheses *before* evaluating the exponent, as usual. Here's an example:

$$(7 - 5)^3$$

First, evaluate $7 - 5$:

$$= 2^3$$

With the parentheses removed, you're ready to evaluate the exponent:

$$= 8$$

Once in a rare while, the exponent itself contains parentheses. As always, evaluate what's in the parentheses first. For example:

$$21^{(19 + \underline{3 \times -6})}$$

This time, the smaller expression inside the parentheses is a mixed-operator expression. I've underlined the part that you need to evaluate first:

$$= 21^{(19 - \underline{18})}$$

Now you can finish off what's inside the parentheses:

$$= 21^1$$

At this point, all that's left is a very simple exponent:

$$= 21$$

So $21^{(19+3\times-6)} = 21$.

Note: Technically, you don't need to put parentheses around the exponent. If you see an expression in the exponent, treat it as though it has parentheses around it. In other words, $21^{19+3\times-6} = 21$ means the same as $21^{(19+3\times-6)} = 21$.

Expressions with nested parentheses

Occasionally, instead of using brackets and parenthesis, an expression has *nested parentheses*, or one or more sets of parentheses inside another set. Here I give you the rule for handling nested parentheses.

When evaluating an expression with nested parentheses, evaluate what's inside the *innermost* set of parentheses first and work your way toward the *outermost* parentheses.

For example, suppose you want to evaluate the following expression:

$$2 + (9 - (\underline{7} - \underline{3}))$$

I underlined the contents of the innermost set of parentheses, so evaluate these contents first:

$$= 2 + (9 - 4)$$

Next, evaluate what's inside the remaining set of parentheses:

$$= 2 + 5$$

Now you can finish things off easily:

$$= 7$$

So $2 + (9 - (7 - 3)) = 7$.

You can follow the same rule when brackets and parenthesis are used. So you can use grouping to simplify $[8 \div (5 - 3)] \times 5$.

$$= [8 \div (5 - 3)] \times 5$$
$$= [8 \div 2] \times 5$$
$$= 4 \times 5$$
$$= 20$$

Putting it all together

In this section, I take you through some trickier examples that combine the order of operations and grouping rules.

The fraction line is a grouping symbol when simplifying. Simplify $\frac{4(7+5)}{2+1}$.

$$\frac{4(7+5)}{2+1}$$

$$\frac{4(12)}{3} = \frac{48}{3} = 16$$

Use both the order of operations and grouping symbols to simplify: $2 + 3^2 (5 - 1)$.

1. **Subtract the 1 from the 5 in the parentheses to get 4.**

   $2 + 3^2 (4)$

2. **Raise the 3 to the second power to get 9.**

   $2 + 9 (4)$

3. **Multiply the 9 and 4 to get 36.**

   $2 + 9 (4) = 2 + 36$

4. **Add to get the final answer.**

   $2 + 36 = 38$

Here's an expression that requires most of the elements from this chapter:

$$4 + (-7 \times (2^{(5-1)} - 4 \times 6))$$

This expression is getting pretty complicated: One set of parentheses containing another set, which contains a third set. To start you off, I underlined what's deep inside this third set of parentheses. This is where you begin evaluating:

$$= 4 + (-7 \times (\underline{2^4} - 4 \times 6))$$

What's left is one set of parentheses inside another set. Again, work from the inside out. The smaller expression here is $2^4 - 4 \times 6$, so evaluate the exponent first, then the multiplication, and finally the subtraction:

$$= 4 + (-7 \times (\underline{16} - 4 \times 6))$$
$$= 4 + (-7 \times (\underline{16 - 24}))$$
$$= 4 + (-7 \times -8)$$

Only one more set of parentheses to go:

$$= 4 + 56$$

At this point, finishing up is easy:

$$= 60$$

Therefore, $4+(-7\times(2^{(5-1)}-4\times6))=60$.

If you'd like more practice, copy the preceding equation down and try solving it step by step with the book closed.

Here's one final example for this chapter, probably about as hard as they come at this stage of maths.

The fraction line and operation all act as grouping symbols.

Simplify: $\dfrac{5[3+(12-2^2)]}{|23-8|}+\dfrac{\sqrt{16-7}}{(-3)^2}$.

1. **Working from the inside out, first square the 2 before subtracting it from the 12. You can also subtract the numbers in the absolute value and the numbers under the radical. Go ahead and square the –3.**

   You can do all these steps at once because none of the results interacts with the others yet

   $$\frac{5[3+(12-4)]}{|15|}+\frac{\sqrt{9}}{9}=\frac{5[3+8]}{|15|}+\frac{\sqrt{9}}{9}$$

2. **Now add the numbers in the brackets and find the square root.**

   $$\frac{5[11]}{15}+\frac{3}{9}$$

3. **Multiply the 5 and 11. Then simplify each of the two fractions by reducing them.**

   $$\frac{55}{15}+\frac{3}{9}=\frac{\overset{11}{\cancel{55}}}{\underset{3}{\cancel{15}}}+\frac{\overset{1}{\cancel{3}}}{\underset{3}{\cancel{9}}}=\frac{11}{3}+\frac{1}{3}$$

4. **You can then add the fractions quite nicely.**

   $$\frac{12}{3}=4$$

Be sure to catch the subtle difference between the two expressions: -2^4 and $(-2)^4$. Simplifying the expression -2^4 you get -16 because the order of operations says to first raise to the fourth power and then apply the negative sign. The expression $(-2)^4 = 16$ because the entire expression in parentheses is raised to the fourth power. This is equivalent to multiplying -2 by itself four times. The multiplication involves an even number of negative signs, so the result is positive (refer to Chapter 3 for the rules on multiplying negative numbers).

In general, if you want a negative number raised to a power, you have to put it in parentheses with the power outside.

Checking Your Answers

Checking your answers when doing algebra is always a good idea, and you can do so on two levels:

- ✔ **Level 1: Does the answer make any sense?** If your bank balance shows $40 million, does that make any sense? Sure, we'd all *like* it to be that but, for most of us, this would be a red flag that something is wrong with our computations.

- ✔ **Level 2: Does actually putting the answer back into the problem give you a true statement? Does it *work?*** This is the more critical check because it gives you more exact information about your answer. The first level helps weed out the obvious errors. This is the final check.

The next sections help you make even more sense of these checks.

Making sense or cents or scents ...

To check whether an answer makes any sense, you have to know something about the topic. A problem will be meaningful if it's about a situation you're familiar with. Just use your common sense. You'll have a good feeling as to whether the money amount in an answer is reasonable.

For example, say you're asked to solve a problem for Jon's weight in kilograms, and your answer is 5. Unless Jon is a dog instead of a person, you probably want to go back and redo the work. Five kilograms (or 5 ounces or 5 tonnes) doesn't make any sense as an answer in this context.

On the other hand, if the problem involves a number of coins in a person's pocket, then five coins seems reasonable. Getting five as the number of goals a player scores in one soccer game may at first seem quite possible but, if you think about it, five goals in one game is a lot. You may want to double-check.

Plugging in to get a charge of your answer

Actually plugging in your answer requires you to go through the manipulations in the problem. You add, subtract, multiply and divide to see if you get a true statement using your answer.

Suppose Jack's mobile phone plan has 400 more minutes than Jill's. If the two of them have a total of 1,400 minutes altogether, how many minutes does Jill have? Does $x = 500$ work for an answer? (I get into algebra in more detail in Parts II and III.) Here's how to check your answer:

1. **Write the problem.**

   Let x represent the number of minutes that Jill has. Jack has $x + 400$ minutes. That means, $x + (x + 400) = 1,400$. The number of minutes Jill has plus the number of minutes Jack has equals 1,400.

2. **Insert the answer into the equation.**

   Replace the variable, x, with your answer of 500 to get $500 + (500 + 400) = 1,400$.

3. **Do the operations and check to see if the answer works.**

   $500 + 900 = 1,400$ is a true statement, so the problem checks. Jill has 500 minutes; Jack has 400 more than that, or 900 minutes; together, they have 1,400 minutes.

You can apply a variation of the preceding steps to check whether $x = 2$ works in the equation $5x [x + 3 (x^2 - 3)] + 1 = 0$.

1. **Write out the equation.**

   $$5x [x + 3 (x^2 - 3)] + 1 = 0$$

2. **Replace the variable with 2.**

   $$5 \times 2 [2 + 3(2^2 - 3)] + 1 = 0$$

3. Do the operations and simplify.

Square the 2 to get $5 \times 2 \, [2 + 3(4 - 3)] + 1 = 0$.

Subtract in the parentheses to get $5 \times 2 \, [2 + 3(1)] + 1 = 0$.

Add in the brackets to get $5 \times 2[5] + 1 = 0$.

Multiply the 5, 2, and 5 to get $50 + 1 \neq 0$.

This time the working does *not* check. You should go back and try again to find a value for x that works.

Part II
Algebra is Part of Everything

Top Five (Okay, Six) Symbols Used in Algebra

- ✓ + means *add* or *find the sum*, *more than* or *increased by*; the result of addition is the *sum*. It also is used to indicate a *positive number*.

- ✓ − means *subtract* or *minus* or *decreased by* or *less than*; the result is the *difference*. It's also used to indicate a *negative number*.

- ✓ × means *multiply* or *times*. The values being multiplied together are the *multipliers* or *factors*; the result is the *product*. Some other symbols meaning *multiply* can be grouping symbols: ( ), [ ], { }, ·, *. In algebra, the × symbol is used infrequently because it can be confused with the variable *x*. The dot is popular because it's easy to write. The grouping symbols are used when you need to contain many terms or a messy expression. By themselves, the grouping symbols don't mean to multiply, but if you put a value in front of a grouping symbol, it means to multiply.

- ✓ ÷ means *divide*. The number that's going into the *dividend* is the *divisor*. The result is the *quotient*. Other signs that indicate division are the fraction line and slash, /.

- ✓ √ means to take the *square root* of something — to find the number, which, multiplied by itself, gives you the number under the sign.

- ✓ π is the Greek letter pi that refers to the irrational number 3.14159 . . . It represents the relationship between the diameter and circumference of a circle.

Part II

Algebra Is Part of Everything

Top Five 'OKay, So?'
Symbols Used in Algebra

In this part . . .

- Start with the basics of algebra to lay the foundations for more advanced aspects.

- Get those variables under control — know how to add, subtract, multiply, divide and distribute them!

- Understand the workings of factorisation and finding the prime factor.

- Work with quadratic expressions and how they can be FOILed (and unFOILed).

Chapter 6

Understanding the Basics of Algebra

In This Chapter

▶ Reminding yourself about the number basics

▶ Working through the algebraic symbols, and communicative and associative properties

▶ Powering up with exponents

▶ Understanding some of the more advanced aspects of exponents

*I*n a nutshell, algebra is a way of generalising arithmetic. Through the use of variables (letters representing numbers) and formulas or equations involving those variables, you solve problems. The problems may be in terms of practical applications, or they may be puzzles for the pure pleasure of the solving. Algebra uses positive and negative numbers, integers, fractions, operations, and symbols to analyse the relationships between values.

In this chapter, you find some of the basics necessary to more easily find your way through the required processes and rules.

Looking at the Basics: Numbers

The different sets of numbers that algebra relies on are important because what they look like and how they behave can set the scene for particular situations or help to solve particular problems.

Algebra uses different sets of numbers, in different circumstances. I describe the different types of numbers here.

Really real numbers

Real numbers are just what the name implies. In contrast to imaginary numbers, they represent *real* values — no pretend or make-believe. Real numbers cover the gamut and can take on any form — fractions or whole numbers, decimal numbers that can go on forever and ever without end, positives and negatives. The variations on the theme are endless.

Counting on natural numbers

A *natural number* (also called a *counting number*) is a number that comes naturally. What numbers did you first use? Remember someone asking, 'How old are you?' You proudly held up four fingers and said, 'Four!' The natural numbers are the numbers starting with 1 and going up by ones: 1, 2, 3, 4, 5, 6, 7 and so on into infinity.

Aha algebra

Dating back to about 2000 BC. with the Babylonians, algebra seems to have developed in slightly different ways in different cultures. The Babylonians were solving three-term quadratic equations, while the Egyptians were more concerned with linear equations. The Hindus made further advances in about the sixth century AD. In the seventh century, Brahmagupta of India provided general solutions to quadratic equations and had interesting takes on 0. The Hindus regarded irrational numbers as actual numbers — although not everybody held to that belief.

The sophisticated communication technology that exists in the world now was not available then, but early civilizations still managed to exchange information over the centuries. In 825 AD, al-Khowarizmi of Baghdad wrote the first algebra textbook. One of the first solutions to an algebra problem, however, is on an Egyptian papyrus that is about 3,500 years old. Known as the Rhind Mathematical Papyrus (after the Scotsman who purchased the 1-foot-wide, 18-foot-long papyrus in Egypt in 1858), the artefact is preserved in the British Museum — with a piece of it in the Brooklyn Museum. Scholars determined that in 1650 BC, the Egyptian scribe Ahmes copied some earlier mathematical works onto the Rhind Mathematical Papyrus.

One of the problems reads, 'Aha, its whole, its seventh, it makes 19.' The *aha* isn't an exclamation. The word *aha* designated the unknown. Can you solve this early Egyptian problem? It would be translated, using current algebra symbols, as: $x + \frac{x}{7} = 19$. The unknown is represented by the x, and the solution is $x = 16\frac{5}{8}$.

It's not hard; it's just messy.

Wholly whole numbers

Whole numbers aren't a whole lot different from natural numbers. Whole numbers are just all the natural numbers plus a 0: 0, 1, 2, 3, 4, 5 and so on into infinity.

Whole numbers act like natural numbers and are used when whole amounts (no fractions) are required. Zero can also indicate none. Algebraic problems often require you to round the answer to the nearest whole number. This makes perfect sense when the problem involves people, cars, animals, houses, or anything that shouldn't be cut into pieces. (I cover whole numbers in Chapter 2.)

Integrating integers

Integers allow you to broaden your horizons a bit. Integers incorporate all the qualities of whole numbers and their opposites (called their *additive inverses*). *Integers* can be described as being positive and negative whole numbers: . . . –3, –2, –1, 0, 1, 2, 3 . . . (Positive and negative numbers are covered in Chapter 3.)

Integers are popular in algebra. When you solve a long, complicated problem and come up with an integer, you can be joyous because your answer is probably right. After all, it's not a fraction! This doesn't mean that answers in algebra can't be fractions or decimals. It's just that most textbooks and reference books try to stick with nice answers to increase the comfort level and avoid confusion. This is my plan in this book, too. After all, who wants a messy answer? (Even though, in real life, that's more often the case.)

Being reasonable: Rational numbers

Rational numbers act rationally! What does that mean? In this case, acting rationally means that the decimal equivalent of the rational number behaves. The decimal ends somewhere, or it has a repeating pattern to it. That's what constitutes 'behaving'.

Some rational numbers have decimals that end such as: 3.4, 5.77623, –4.5. Other rational numbers have decimals that repeat the same pattern, such as 3.164164164, or 0.666666666. A horizontal bar over the 164 and the 6 lets you know that these numbers repeat forever.

In *all* cases, rational numbers can be written as fractions. Each rational number has a fraction that it's equal to. So one definition of a *rational number* is any number that can be written as a fraction, 3.164164$\overline{164}$, where *p* and *q* are integers (except *q* can't be 0). If a number can't be written as a fraction, it isn't a rational number. Rational numbers appear in Chapter 11, where you see quadratic equations, and in Part III, where the applications are presented.

Restraining irrational numbers

Irrational numbers are just what you may expect from their name — the opposite of rational numbers. An *irrational number* cannot be written as a fraction, and decimal values for irrationals never end and never have a nice pattern to them. Whew! Talk about irrational! For example, pi, with its never-ending decimal places, is irrational. Irrational numbers are often created when using the quadratic formula, as you see in Chapter 11.

Picking out primes and composites

A number is considered to be prime if it can be divided evenly only by 1 and by itself. The first *prime numbers* are: 2, 3, 5, 7, 11, 13, 17, 19, 23, 29, 31 and so on. The only prime number that's even is 2, the first prime number. Mathematicians have been studying prime numbers for centuries, and prime numbers have them stumped. No-one has ever found a formula for producing all the primes. Mathematicians just assume that prime numbers go on forever.

A number is *composite* if it isn't prime — if it can be divided by at least one number other than 1 and itself. So the number 12 is composite because it's divisible by 1, 2, 3, 4, 6 and 12.

Deciphering the Symbols in Algebra Operations

In Chapter 1, I cover some of the basic terms used in algebra. The basics of algebra also involve symbols. Algebra uses symbols for quantities, operations, relations or grouping. The symbols are shorthand and are much more efficient than writing out the words or meanings. But you need to know what the symbols represent, and the following list shares some of that info:

- ✔ + means *add* or *find the sum, more than* or *increased by*; the result of addition is the *sum*. It also is used to indicate a *positive number*.

- ✔ – means *subtract* or *minus* or *decreased by* or *less than*; the result is the *difference*. It's also used to indicate a *negative number*.

- ✔ × means *multiply* or *times*. The values being multiplied together are the *multipliers* or *factors*; the result is the *product*. Some other symbols meaning *multiply* can be grouping symbols: ( ), [ ], { }, ·, *. In algebra, the × symbol is used infrequently because it can be confused with the variable *x*. The dot is popular because it's easy to write. The grouping symbols are used when you need to contain many terms or a messy expression. By themselves, the grouping symbols don't mean to multiply, but if you put a value in front of a grouping symbol, it means to multiply.

- ✔ ÷ means *divide*. The number that's going into the *dividend* is the *divisor*. The result is the *quotient*. Other signs that indicate division are the fraction line and slash, /.

- ✔ √ means to take the *square root* of something — to find the number, which, multiplied by itself, gives you the number under the sign.

- ✔ π is the Greek letter pi that refers to the irrational number: 3.14159 … It represents the relationship between the diameter and circumference of a circle.

Grouping

When a car manufacturer puts together a car, several different things have to be done first. The engine experts have to construct the engine with all its parts. The body of the car has to be mounted onto the chassis and secured, too. Other car specialists have to perform the tasks that they specialise in as well. When these tasks are all accomplished in order, the car can be put together. The same thing is true in algebra. You have to do what's inside the *grouping* symbol before you can use the result in the rest of the equation.

Grouping symbols tell you that you have to deal with the *terms* inside the grouping symbols *before* you deal with the larger problem. If the problem contains grouped items, do what's inside a grouping symbol first, and then follow the order of operations. (Refer to Chapter 5 for more on order of operations and grouping symbols.)

Even though the order of operations and grouping-symbol rules are fairly straightforward, it's hard to describe, in words, all the situations that can come up in these problems. The examples in Chapter 5 should clear up any questions you may have.

Defining relationships

Algebra is all about relationships — not the he-loves-me-he-loves-me-not kind of relationship — but the relationships between numbers or among the terms of an equation. Although algebraic relationships can be just as complicated as romantic ones, you have a better chance of understanding an algebraic relationship. The symbols for the relationships are given here:

- ✔ = means that the first value is *equal to* or the same as the value that follows.

- ✔ ≠ means that the first value *is not equal to* the value that follows.

- ✔ ≈ means that one value is *approximately the same* or *about the same* as the value that follows; this is used when rounding numbers.

- ✔ ≤ means that the first value is *less than or equal to* the value that follows.

- ✔ < means that the first value is *less than* the value that follows.

- ✔ ≥ means that the first value is *greater than or equal to* the value that follows.

- ✔ > means that the first value is *greater than* the value that follows.

Taking on algebraic tasks

Algebra involves symbols, such as variables and operation signs, which are the tools that you can use to make algebraic expressions more usable and readable. These things go hand in hand with simplifying, factoring and solving problems, which are easier to solve if broken down into basic parts. Using symbols is actually much easier than wading through a bunch of words.

Here's what I mean by simplifying, expanding, factoring and solving:

- ✔ To *simplify* means to combine all that can be combined, cut down on the number of terms, and put an expression in an easily understandable form.

✔ To *expand* means to 'get rid of' brackets and parentheses. When expanding it is important to remember that everything in the bracket is affected by the *pronumeral* or number outside the bracket.

✔ To *factor* means to change two or more terms to just one term. (See Chapter 8 for more on factoring.)

✔ To *solve* means to find the answer. In algebra, it means to figure out what the variable stands for. (You can find more on solving equations in Part III.)

Equation solving is fun because it has a point. You solve for something (often a variable, such as *x*) and get an answer that you can check to see whether you're right or wrong. It's like a puzzle. It's enough for some people to say, 'Give me an *x*.' What more could you want? But solving these equations is just a means to an end. The real beauty of algebra shines when you solve some problem in real life — a practical application. Are you ready for these two words: *story problems*? Story problems are the whole point of doing algebra. Why do algebra unless there's a good reason? Oh, I'm sorry — you may just like to solve algebra equations for the fun alone. (Yes, some folks are like that.) But other folks love to see the way a complicated paragraph in the English language can be turned into a neat, concise expression, such as, 'The answer is three bananas.'

Going through each step and using each tool to play this game is entirely possible. *Simplify, factor, solve, check.* That's good! Lucky you. It's time to dig in!

Associating and Commuting with Expressions

Algebra operations follow certain rules, and those rules have certain properties. The properties usually make computations easier. In this section, I talk about two of those properties — the commutative property and the associative property.

Reordering operations: The commutative property

Before discussing the commutative property, take a look at the word *commute*. You probably commute to school and know that whether you're travelling from home to school or from school to home, the distance is the same: The distance doesn't change because you change directions (although getting home when you're starving may make that distance *seem* longer).

The same principle is true of *some* algebraic operations: It doesn't matter whether you add $1 + 2$ or $2 + 1$, the answer is still 3. Likewise, multiplying 2×3 or 3×2 yields 6.

The *commutative property* means that you can change the order of the numbers in an operation without affecting the result. Addition and multiplication are commutative. Subtraction and division are not. So:

$$a + b = b + a$$

$$a \times b = b \times a$$

$$a - b \neq b - a \text{ (except in a few special cases)}$$

$$a \div b \neq b \div a \text{ (except in a few special cases)}$$

In general, subtraction and division are *not* commutative. The special cases occur when you choose the numbers carefully. For example, if a and b are the same number, the subtraction appears to be commutative because switching the order doesn't change the answer. In the case of division, if a and b are opposites, you get -1 no matter which order you divide them in.

Take a look at how the commutative property works:

- $4 + 5 = 9$ and $5 + 4 = 9$, so $4 + 5 = 5 + 4$.
- $3 \times (-7) = -21$ and $(-7) \times 3 = -21$, so $3 \times (-7) = (-7) \times 3$.
- $(-5) - (+2) = -7$ and $(+2) - (-5) = +7$, so $(-5) - (+2) \neq (+2) - (-5)$.
- $(-6) \div (+1) = -6$ and $(+1) \div (-6) = -\dfrac{1}{6}$ so $(-6) \div (+1) \neq (+1) \div (-6)$.

Associating expressions: The associative property

The commutative property has to do with the order of the numbers when you perform an operation. The associative property has to do with how the numbers are grouped when you perform operations on more than two numbers.

Think about what the word *associate* means. When you associate with someone, you're close to the person, or you're in the same group with the person. Say that Anika, Becky and Cora associate. Whether Anika walks over to pick up Becky and the two of them walk to Cora's and pick her up, or Cora is at Becky's house and Anika walks over and picks up both of them at the same time, the same result occurs — the three girls are all walking together at the end.

The *associative property* means that even if the grouping of the operation changes, the result remains the same. (If you need a reminder about grouping, check out Chapter 5.) Addition and multiplication are associative. Subtraction and division are *not* associative operations. So:

$$a + (b + c) = (a + b) + c$$

$$a \times (b \times c) = (a \times b) \times c$$

$$a - (b - c) \neq (a - b) - c \text{ (except in a few special cases)}$$

$$a \div (b \div c) \neq (a \div b) \div c \text{ (except in a few special cases)}$$

Here's how the associative property works:

- $4 + (5 + 8) = 4 + 13 = 17$ and $(4 + 5) + 8 = 9 + 8 = 17$, so $4 + (5 + 8) = (4 + 5) + 8$

- $3 \times (2 \times 5) = 3 \times 10 = 30$ and $(3 \times 2) \times 5 = 6 \times 5 = 30$, so $3 \times (2 \times 5) = (3 \times 2) \times 5$

- $13 - (8 - 2) = 13 - 6 = 7$ and $(13 - 8) - 2 = 5 - 2 = 3$, so $13 - (8 - 2) \neq (13 - 8) - 2$

- $48 \div (16 \div 2) = 48 \div 8 = 6$ and $(48 \div 16) \div 2 = 1\frac{1}{2}$ so $48 \div (16 \div 2) \neq (48 \div 16) \div 2$

The commutative and associative properties come in handy when you work with algebraic expressions. You can change the order of some numbers or change the grouping to make the work less messy or more convenient. Just keep in mind that you can commute and associate addition and multiplication operations, but not subtraction or division.

You can use the commutative and associative properties to find the answer to the problem: $417 + 932 + (-416) + (-432) + 800$. To simplify the expression, you would normally just move from left to right, adding and subtracting in order, but rearranging the numbers works better. Use the commutative property to switch the 932 and −416:

$$417 + (-416) + 932 + (-432) + 800$$

Now group (associate) the first two numbers and the third and fourth numbers. Combine them and add up the results:

$$(417 + [-416]) + (932 + [-432]) + 800 = 1 + 500 + 800 = 1{,}301$$

You can do the computation in your head when you use these handy properties.

I Got the Power! Using Exponents

In Part I of this book, I talk about the four main operations — addition, subtraction, multiplication and division. In algebra, a fifth operation is important — *exponentiation*, which you can think of as repeated multiplication. I touch on exponents in Chapter 5, where I cover the order of operations and I expand on them here.

Understanding what exponents are

Exponents, also known as *powers* or *indices*, are those small symbols, slightly higher and to the right of numbers, and were developed so that mathematicians wouldn't have to keep repeating themselves! What is an exponent? An *exponent* is the small, superscripted number to the upper right of the larger number that tells you how many times you multiply the larger number, called the *base*. That is, three to the fourth power (3^4) is 3 multiplied 4 times. Got that? Now, here's what happens:

$$3^4 = 3 \times 3 \times 3 \times 3 = [(3 \times 3) \times 3] \times 3 = 81$$

So, really, three to the fourth power (3^4) is another way of saying 81.

To help you understand their significance, allow me to give you an analogy.

You know that $4 \times 3 = 12$, because there are 12 things in 4 groups of 3. If you didn't *know* the product 4×3, you can find it in several ways. You could lay out 4 groups of 3 things and count them one by one, for example. Or you could use the associative property of multiplication (covered in the preceding section) to think of 4×3 as $2 \times (2 \times 3)$, so 4×3 is twice as much 2×3. Finally, you could think $4 \times 3 = 3 + 3 + 3 + 3$. That is, one way to compute products is by using repeated addition.

It's the same with exponents: $4^3 = 64$. From 4 and 3, you compute a third number, 64. Just as you *can* compute 4×3 by using repeated addition, you can compute 4^3 using repeated multiplication: $4^3 = 4 \times 4 \times 4$. But there are other ways too, and these ways depend on properties of exponentiation as an operation. You can double 2×3 to get 4×3 using the associative property of multiplication, and properties of exponentiation allow you to relate 4^3 to 2^3. These properties are known as *rules for operating with exponents*.

Exponents appear in Years 9 and 10. In the following statements, A is presumed to be a positive number:

$$A^B \times A^C = A^{B+C}$$

$$\frac{A^D}{A^E} = A^{D-E}$$

$$A^0 = 1$$

$$\left(A^B\right)^C = A^{B \times C}$$

$$\left(A^B B^X\right)^C = A^{B \times C} B^{X \times C}$$

$$\frac{\left(A^M\right)^N}{\left(B^X\right)^Y} = \frac{A^{MN}}{B^{YX}}$$

$$A^{-B} = \frac{1}{A^B}$$

$$\sqrt{A} = A^{\frac{1}{2}}$$

More detail about each specific law is covered in the next section, and a couple more are added. You can understand these rules better by way of examples. You can see the first rule, that $A^B \times A^C = A^{B+C}$, by thinking of $3^2 \times 3^4 = (3 \times 3) \times (3 \times 3 \times 3 \times 3) = 3^6$. Six threes are multiplied. The second rule you can see by thinking about $(3^2)^4 = 3^2 \times 3^2 \times 3^2 \times 3^2 = 3^8$. Eight threes are multiplied together. The fourth rule is the logical consequence of the first rule, and the third rule that $A^0 = 1$ when A is any positive number. Here's why: $3^2 \times 3^{-2} = 3^0$ by the first rule. Then $3^0 = 1$, so 3^{-2} has to be the reciprocal of 3^2.

Each of these rules is useful going in both directions. You don't have to view these equations as machines that transform the left side into the right side. Instead, each side of each equation has the same value as the other side. Sometimes you have something that looks like this: $A^B \times A^C$, and it's useful to write it as A^{B+C}. Sometimes it goes the other way around. What matters is the equivalence — or *sameness* — of both sides of each equation.

The first index law

When algebra was first written with symbols — instead of with all words — there were no exponents. If you wanted to multiply the variable y times itself six times, you'd write it: $yyyyyy$. (Kinda like a 3-year-old asking: 'Why, why,

why, why, why, why?') Writing the *variable* (the letter representing a number) over and over can get tiresome (just like 3-year-olds), so the wonderful system of exponents was developed.

Writing numbers with exponents is one thing — knowing what these exponents mean and what you can do with them is another thing altogether. Using exponents is so convenient that it's worth the time and trouble to find out the rules for using them correctly.

The base of an exponential expression can be any *real number*. (Real numbers are the rational and irrational numbers together — refer to the earlier section 'Looking at the Basics: Numbers'.) The exponent (the power) can be any real number, too. An exponent can be positive, negative or fractional. What power!

The symbol $\cdot$ means multiply (as does $\times$). I've used $\cdot$ in this chapter and throughout the rest of the book, where needed, to avoid any confusion with the algebraic use of x.

When a number x is involved in repeated multiplication of x times itself, the number n can be used to describe how many multiplications are involved: $x^n = x \cdot x \cdot x \cdot x \cdot x \ldots n$ times.

Even though the x in the expression x^2 can be any real number and the n can be any real number, they can't both be 0 at the same time. For example, 0^0 really has no meaning in algebra. It takes a calculus course to prove why this restriction is so. Also, if x is equal to 0, then n can't be negative.

Here are some examples using exponential notation:

- $2^4 = 2 \times 2 \times 2 \times 2 = 16$
- $3^5 = 3 \times 3 \times 3 \times 3 \times 3 = 243$
- $10^8 = 10 \times 10 \times 10 \times 10 \times 10 \times 10 \times 10 \times 10 = 100{,}000{,}000$

(See the 'Working with negative exponents' section, later in the chapter, for more on the last example.)

To see the benefits of using exponents, write the expression $3^3x^2y^4z^6$ without exponents. In this example, several bases are multiplied together. Each base has its own, separate exponent. The x, y and z are variables representing real numbers:

$$3^3x^2y^4z^6 = 3 \cdot 3 \cdot 3 \cdot x \cdot x \cdot y \cdot y \cdot y \cdot y \cdot z \cdot z \cdot z \cdot z \cdot z \cdot z$$

$$3x^2y^2 = 3 \cdot x \cdot x \cdot y \cdot y$$

$$3a^6 \times 4b^2 = 3 \cdot 4 \cdot a \cdot a \cdot a \cdot a \cdot a \cdot a \cdot b \cdot b$$

You can see why using the powers is preferable. And in the next example, the base is a binomial.

The expression $(a + b)^3$ without an exponent is as follows:

$$(a + b)^3 = (a + b) \times (a + b) \times (a + b)$$

The parentheses mean that you add the two values together before applying the exponent.

You can multiply many exponential expressions together without having to change their form into the big or small numbers they represent. The only requirement is that the bases of the exponential expressions that you're multiplying have to be the same. The answer is then a nice, neat exponential expression.

You *can* multiply $2^4 \times 2^6$ and $a^5 \times a^8$, but you *cannot* multiply $3^6 \times 4^7$ because the bases are not the same. To multiply powers of the same base, add the exponents together: $x^a \times x^b = x^{a + b}$.

Here are some examples of finding the products of the numbers by adding the exponents:

- $2^4 \times 2^9 = 2^{4 + 9} = 2^{13}$
- $a^5 \times a^8 = a^{13}$
- $4^a \times 4^2 = 4^{a + 2}$

Often, you find algebraic expressions with a whole string of factors; you want to simplify the expression, if possible. When more than one base is used in an expression with powers, you combine the numbers with the same bases, find the values, and then write them all together.

Here's how to simplify the following expressions:

- $3^2 \times 2^2 \times 3^3 \times 2^4 = 3^{2 + 3} \times 2^{2 + 4} = 3^5 \times 2^6$: The two factors with base 3 combine, as do the two factors with base 2.

- $4x^6y^5x^4y = 4x^{6 + 4}y^{5 + 1} = 4x^{10}y^6$: The number 4 is a coefficient, which is written before the rest of the factors.

When no exponent is showing, such as with y in the preceding example, you assume that the exponent is 1. In the preceding example, you see that the factor y was written as y^1 so its exponent could be added to that in the other y factor.

You can add exponents when multiplying numbers with the same base. But you can multiply numbers that have the same power (in a multiplication problem). The rule is that: $a^n \times b^n = (a \times b)^n$. So, if you have $4^8 \times 7^8$, you simplify it to $(4 \times 7)^8 = 28^8$. You'd rather leave the simplified expression as the power of 28, because the actual number is huge!

The second index law

You can divide exponential expressions, leaving the answers as exponential expressions, as long as the bases are the same. Division is the opposite of multiplication, so it makes sense that, because you add exponents when multiplying numbers with the same base, you *subtract* the exponents when dividing numbers with the same base. Easy enough?

To divide powers with the same base, subtract the exponents:

$$\frac{x^a}{x^b} = x^a \div x^b = x^{a-b}$$

where x can be any real number except 0. (***Remember***: You can't divide by 0.)

Here are some examples of simplifying expressions by dividing:

- $2^{10} \div 2^4 = 2^{10-4} = 2^6$: These exponentials represent the problem $1{,}024 \div 16 = 64$. It's much easier to leave the numbers as bases with exponents.
- $\frac{4x^6y^3z^2}{2x^2y^3z} = 2x^{6-2}y^{3-3}z^{2-1} = 2x^4y^0z^1 = 2x^2z$: The variables represent numbers, so writing this out the long way would be

$$\frac{2 \cdot 2 \cdot x \cdot x \cdot x \cdot x \cdot x \cdot x \cdot y \cdot y \cdot y \cdot z \cdot z}{2 \cdot x \cdot x \cdot x \cdot x \cdot y \cdot y \cdot y \cdot z}$$

$$= \frac{2 \cdot 2 \cdot x \cdot x \cdot x \cdot x \cdot x \cdot x \cdot y \cdot y \cdot y \cdot z \cdot z}{2 \cdot x \cdot x \cdot x \cdot x \cdot y \cdot y \cdot y \cdot y \cdot z}$$

By crossing out the common factors, all that's left is $2x^2z$.

The second index law can also be utilised when working with algebraic expressions — for example, $9x^3y \div 6xy^2$.

Now, believe it or not, the easiest way to solve this kind of problem is to set it out as a fraction. I know, I know — I can hear you sigh — but, believe me, it is the easiest way to ensure that you don't make an error.

Here's the example set as a fraction.

$$= \frac{9 \times x^3 \times y}{6 \times x \times y^2}$$

Now expand the numerator and the denominator.

$$= \frac{9 \times x \times x \times x \times x \times y}{6 \times x \times x \times y \times y}$$

The next bit is the best part because you get to cross things out — and, let's face it, that is always fun! Cross out as many xs on top and on the bottom as match. Repeat this for the ys. Remember to also simplify the numbers if you can.

$$= \frac{9\!\!\!/^3 \times \not x \times x \times x \times x \times y}{6\!\!\!/^2 \times \not x \times y \times y}$$

Now you can rewrite the fraction showing only what is left.

$$= \frac{3 \times x \times x}{2 \times y}$$

Rewrite the answer using exponents. To work out the number of exponents you have, simply count the letters, remembering that if only one is left, you don't need to write the 1.

$$= \frac{3x^2}{2y}$$

This method is foolproof, and you can also get a little creative and add some colour, using your highlighters to cross things out, making sure that you have the same number on the top and the bottom.

The following example shows the workings for a slightly more complex question, so that you can see that this method always works.

$$15a^3b^4c^2 \div 5ab^2c = \frac{15 \times a^3 \times b^4 \times c^2}{5 \times a \times b^2 \times c}$$

$$= \frac{15 \times a \times a \times a \times b \times b \times b \times b \times c \times c}{5 \times a \times b \times b \times c}$$

$$= \frac{15\!\!\!/^3 \times \not a \times a \times a \times \not b \times \not b \times b \times b \times \not c \times c}{5\!\!\!/^1 \times \not a \times \not b \times \not b \times \not c}$$

$$= \frac{3 \times a \times a \times b \times b \times c}{1}$$

$$= \frac{3a^2b^2c}{1}$$

$$= 3a^2b^2c$$

Need I say more? Well, yes, there's lots more to say — see the following section.

Getting Complicated with Exponents

You may be confident when working with the basics of exponents, and multiplying and dividing exponents as needed. But what happens when you need to work with zero, negative exponents, powers of powers and square roots? The following sections provide all the answers.

The third index law: The power of zero

If x^3 means $x \cdot x \cdot x$, what does x^0 mean? Well, it doesn't mean x times 0, so the answer isn't 0. x represents some unknown real number; real numbers can be raised to the 0 power — except that the base just can't be 0. To understand how this works, use the following rule for division of exponential expressions involving 0.

Any number to the power of 0 equals 1 as long as the base number is not 0. In other words, $a^0 = 1$ as long as $a \neq 0$.

Consider the situation where you divide 2^4 by 2^4 by using the rule for dividing exponential expressions, which says that if the base is the same, subtract the two exponents in the order that they're given. Doing this you find that the answer is $2^{4-4} = 2^0$. But $2^4 = 16$, so $2^4 \div 2^4 = 16 \div 16 = 1$. That means that $2^0 = 1$. This is true of all numbers that can be written as a division problem, which means that it's true for all numbers except those with a base of 0.

Here are some examples of simplifying, using the rule that when you raise a real number a to the 0 power, you get 1:

- $m^2 \div m^2 = m^{2-2} = m^0 = 1$.

- $4x^3y^4z^7 \div 2x^3y^3z^7 = 2x^{3-3}y^{4-3}z^{7-7} = 2x^0y^1z^0 = 2y$. Both x and z end up with exponents of 0, so those factors become 1. Neither x nor z may be equal to 0.

- $\dfrac{(2x^2+3x)^4}{(2x^2+3x)^4} = (2x^2+3x)^{4-4} = (2x^2+3x)^0 = 1$.

The fourth, fifth and sixth index laws: Powers of powers

Because exponents are symbols for repeated multiplication, one way to write $(x^3)^6$ is $x^3 \cdot x^3 \cdot x^3 \cdot x^3 \cdot x^3 \cdot x^3$. Using the multiplication rule, where you just add all the exponents together, you get $x^{3+3+3+3+3+3} = x^{18}$. Wouldn't it be just grand if the rule for raising a power to a power was just to multiply the two exponents together? Lucky you!

To raise a power to a power, use this formula: $(x^n)^m = x^{nm}$. In other words, when the whole expression, x^n, is raised to the mth power, the new power of x is determined by multiplying n and m together.

This formula is known as the fourth index law, which states $(a^m)^n = a^{mn}$.

The fifth index law states that $(a^m b^p)^n = a^{mn} b^{pn}$.

The sixth law states that $\left(\frac{a^m}{b^m}\right)^n = \frac{a^{mn}}{b^{pn}}$.

These laws can be remembered by thinking about expanding brackets; that is; everything inside the bracket is affected by the index outside the bracket.

Here are some examples of simplifying using the rule for raising a power to a power:

✔ $(x^2)^3 = x^{(2)(3)} = x^6$

✔ $(3x^2y^3)^2 = 3^2 x^{2 \times 2} y^{3 \times 2} = 9x^4y^6$: Each factor in the parentheses is raised to the power outside the parentheses.

The following are some further examples and their workings. Notice that the order of operations is observed as required.

$$(x^2)^7 = x^{2\times 7}$$
$$= x^{14}$$
$$(4b^2)^2 = 4^2 \times b^{2\times 2}$$
$$= 16b^4$$
$$(m^4 n^2)^6 = m^{4\times 6} \times n^{2\times 6}$$
$$= m^{24} n^{12}$$
$$(3a^5 b)^2 = 3^2 \times a^{5\times 2} \times b^{1\times 2}$$
$$= 9a^{10} b^2$$
$$\left(\frac{r^6}{p^2}\right)^3 = \frac{r^{6\times 3}}{p^{2\times 3}}$$
$$= \frac{r^{18}}{p^6}$$
$$\left(\frac{2t^3}{u}\right)^4 = \frac{2^4 \times t^{3\times 4}}{u^{1\times 4}}$$
$$= \frac{16t^{12}}{u^4}$$

The seventh index law: Working with negative exponents

Negative exponents are a neat little creation. They mean something very specific and have to be handled with care, but they are oh, so convenient to have.

You can use a negative exponent to write a fraction without writing a fraction! This is because negative exponents are a way of writing powers of fractions or decimals without using the fraction or decimal. For example, instead of writing $\left(\frac{1}{10}\right)^{14}$, you can write 10^{-14}.

Using negative exponents is also a way to combine expressions with the same base, whether the different factors are in the numerator or denominator. It's a way to change division problems into multiplication problems.

A *reciprocal* of a number is the multiplicative inverse of the number. The product of a number and its reciprocal is equal to 1.

The reciprocal of x^a is $\frac{1}{x^a}$, which can be written as x^{-a}. The variable x is any real number except 0, and a is any real number. Also, to get rid of the negative exponent, you write: $x^{-a} = \frac{1}{x^a}$.

Here are some examples of changing numbers with negative exponents to fractions with positive exponents:

- $2^{-3} = \frac{1}{2^3} = \frac{1}{8}$. The reciprocal of 2^3 is $\frac{1}{2^3} = 2^{-3}$.

- $z^{-4} = \frac{1}{z^4}$. The reciprocal of z^4 is $\frac{1}{z^4} = z^{-4}$. In this case, z cannot be 0.

- $6^{-1} = \frac{1}{6}$. The reciprocal of 6 is $\frac{1}{6} = 6^{-1}$.

But what if you start out with a negative exponent in the denominator? What happens then? Look at the fraction $\frac{1}{3^{-4}}$. If you write the denominator as a fraction, you get $\frac{1}{\frac{1}{3^4}}$. Then, changing the *complex fraction* (a fraction with a fraction in it) to a division problem: $\frac{1}{\frac{1}{3^4}} = 1 \div \frac{1}{3^4} = 1 \cdot \frac{3^4}{1} = 3^4$. So, to simplify a fraction with a negative exponent in the denominator, you can do a switcheroo: $\frac{1}{3^{-4}} = 3^4$.

Here are some examples of simplifying the fractions by getting rid of the negative exponents:

$$\frac{x^2 y^3}{3z^{-4}} = \frac{x^2 y^3 z^4}{3}$$

$$\frac{4a^3 b^5 c^6 d}{a^{-1} b^{-2}} = 4a^3 a^1 b^5 b^2 c^6 d = 4a^4 b^7 c^6 d$$

The eighth index law

When you do square roots, the symbol for that operation is a radical, $\sqrt{\ }$. A cube root has a small 3 in front of the radical; a fourth root has a small 4, and so on.

The radical is a non-binary operation (involving just one number) that asks you, 'What number times itself gives you this number under the radical?' Another way of saying this is: 'If $\sqrt{\ }$, then $b^2 = a$.' Alternately, the index law can be applied:

$$\sqrt{4} = 4^{\frac{1}{2}} = 2$$

$$\sqrt[3]{8} = 8^{\frac{1}{3}} = 2$$

$$\sqrt[4]{16} = 8^{\frac{1}{4}} = 2$$

Finding square roots is a relatively common operation in algebra, but working with and combining the roots isn't always so clear.

Expressions with radicals can be multiplied or divided as long as the root power *or* the value under the radical is the same. Expressions with radicals cannot be added or subtracted unless *both* the root power *and* the value under the radical are the same.

Here are some examples of simplifying the radical expressions when possible:

- $\sqrt{2} \cdot \sqrt{3} = \sqrt{6}$: These *can* be combined because it's multiplication, and the root power *is* the same.

- $\sqrt{8} \div \sqrt{4} = \sqrt{2}$: These *cannot* be combined because it's addition, and the value under the radical is *not* the same.

- $\sqrt{2} - \sqrt{3}$: These *cannot* be combined because it's subtraction, and the value under the radical is *not* the same.

- $\sqrt{7} - \sqrt[3]{4}$: These *cannot* be combined because it's subtraction, and the root power *isn't* the same.

- $2\sqrt[3]{3} - \sqrt[3]{3}$: These *can* be combined because the root power and the numbers under the radical *are* the same.

When the numbers inside the radical are the same, you can see some nice combinations involving addition and subtraction. Multiplication and division can be performed whether they're the same or not. The *root power* refers to square root, cube root, fourth root, and so on.

Here are the rules for adding and subtracting radical expressions. Assume that a and b are positive values.

- ✓ $4\sqrt{3} + 2\sqrt{3} = 6\sqrt{3}$: Addition and subtraction can be performed if the root power and value under the radical are the same.

- ✓ $\sqrt{5} + 3\sqrt{5} = 4\sqrt{5}$: Addition and subtraction can be performed if the root powers are the same.

- ✓ $\sqrt[3]{7} + 4\sqrt[3]{7} + 2\sqrt[3]{7} = 7\sqrt[3]{7}$: Addition and subtraction can be performed if the root powers are the same.

- ✓ $\sqrt[4]{5} + 11\sqrt[4]{5} + 4\sqrt[4]{5} = 16\sqrt[4]{5}$: Addition and subtraction can be performed if the root powers are the same.

- ✓ $m\sqrt{a} + n\sqrt{a} = (m+n)\sqrt{a}$: Addition and subtraction can be performed if the root powers are the same.

- ✓ $m\sqrt{a} - n\sqrt{a} = (m-n)\sqrt{a}$: Addition and subtraction can be performed if the root powers are the same.

Notice that the square root of a 1 followed by an even number of zeros is always a 1 followed by half that many zeros.

The convention that mathematicians have adopted is to use fractions in the powers to indicate that this stands for a root or a radical. The fractional exponents are easier to use when combining factors, and they're easier to type — for example, $\sqrt{a} \times \sqrt{a} = \sqrt{a^2}$.

Notice that, when no number is outside and to the upper left of the radical, you assume that it's a 2, for a square root. Also, recall that when raising a power to a power, you multiply the exponents (refer to the preceding section).

When changing from radical form to fractional exponents:

- ✓ $\sqrt{a}\sqrt{b} = \sqrt{ab}$: The nth root of a can be written as a fractional exponent with a raised to the reciprocal of that power.

- ✓ $\dfrac{\sqrt{a}}{\sqrt{b}} = \sqrt{\dfrac{a}{b}}$: When the nth root of a^m is taken, it's raised to the $\sqrt{1} = 1$th power.

This rule involving changing radicals to fraction exponents allows you to simplify the following expressions. Note that when using the 'Powers of powers' rule, the bases still have to be the same.

Here are some examples of simplifying each expression, combining like factors:

- ✔ $\sqrt{9} = 3$

- ✔ $\sqrt{16} = 4$: Leave the exponent as $\sqrt{25} = 5$. Don't write the exponent as a mixed number.

- ✔ $\sqrt{36} = 6$: The exponents can't really be combined, because the bases are not the same.

Comparing with Exponents

Comparing amounts is easier when you use exponents. This is because the nice thing about powers of 10 is that the power tells you how many zeros are in the answer.

Try to compare these two numbers: 943,260,000,000,000,000,000,000 and 8,720,000,000,000,000,000,000,000. Which is bigger? The first number may *look* bigger because of the first three digits, but this is deceiving. To compare large numbers, rewrite them as products involving exponents.

Using the previous numbers, you can write them as follows (with just one numeral before the decimal point):

$$943{,}260{,}000{,}000{,}000{,}000{,}000{,}000 =$$

$$9.4326 \times 100{,}000{,}000{,}000{,}000{,}000{,}000{,}000 =$$

$$9.4326 \times 10^{23}$$

and

$$8{,}720{,}000{,}000{,}000{,}000{,}000{,}000{,}000 =$$

$$8.72 \times 1{,}000{,}000{,}000{,}000{,}000{,}000{,}000{,}000 =$$

$$8.72 \times 10^{24}$$

The number with the higher power of 10 is the larger number:

$$8.72 \times 10^{24} > 9.4326 \times 10^{23}$$

If the powers are the same, compare the numbers multiplying the power of 10.

Why is the number with the higher power of 10 larger? Look at these two numbers that are a little more manageable (they don't have over 20 zeros): 3×10^2 and 9×10^1. That's comparing $3 \times 100 = 300$ with $9 \times 10 = 90$. Even though the 9 is bigger than the 3, it's the larger power of 10 that 'wins'.

Taking notes on scientific notation

When people talk about distances between planets, the number of grains of sand or the amount of money spent by the government, they have to use very large numbers. When the topic turns to measurements of plant or animal cells, the size of atoms or other such teeny things, they use very small numbers. *Scientific notation* is a standard way of recording these very large and very small numbers so they can fit on one line in the page of a book and so they can be compared more easily. Computations with large and small numbers are easier in scientific-notation form, too.

For example, you write 3,456 as 3.456×10^3 in scientific notation. Doing so is especially useful for working with very large numbers (such as 1.52×10^{11}, which is approximately the number of metres you are from the sun right now), and with very small numbers (such as 1.2×10^{-8}, which is approximately the diameter of a virus in metres). Scientific notation makes it much easier to pay attention to the size of these numbers than writing out all those zeroes. When students compute with numbers in scientific notation, the rules for operating with exponents simplify these computations greatly.

The form for a number written in scientific notation is: $N \times 10^a$, where N is a number between 1 and 9 (you don't use 10 because it has two digits), and where a is an *integer* (a positive or negative number).

You can write large and small numbers in scientific notation by moving the decimal point until you create a number from 1 up to 10, and then indicating how many places the decimal point was moved by the power you raise 10 to. Whether the power of 10 is positive or negative depends on whether you move the decimal to the right or to the left: Moving the decimal to the right makes the exponent negative; moving it to the left makes the exponent positive.

For example, the star Rigel, in the constellation Orion, is 777 light years away. A *light year* is the distance that light travels in a year. So, if light travels at 186,000 miles per second, you multiply: 186,000 times 60 seconds in a minute times 60 minutes in an hour times 24 hours in a day times 365 days in a year times 777 years to get about 4,557,645,792,000,000 miles. Written exponentially, the distance to Rigel is about 4.558×10^{15} miles away.

I've already mentioned another example — viruses, which may be the smallest microbes, but they sure pack a nasty wallop! The Ebola virus measures $\frac{1}{25,000}$ inch, or 0.00004 inch. The influenza virus measures $\frac{5}{1,000,000}$ inch, or 0.000005 inch. In scientific notation, the two viruses measure 4×10^{-5} and 5×10^{-6} inch, respectively.

To write a number in scientific notation:

1. **Determine where the decimal point is in the number and move it left or right until you have exactly one digit to the left of the decimal point.**

   This gives you a number between 1 and 9.

2. **Count how many places (digits) you had to move the decimal point from its original position.**

   This is the absolute value of your exponent.

3. **If you moved the original decimal point to the left, your exponent is positive. If you moved the original decimal point to the right, your exponent is negative.**

4. **Rewrite the number in scientific notation by making a product of your new, between-1-and-9 number, times a 10 raised to the power of your exponent.**

Here are some numbers written in scientific notation:

- ✔ **$41,000 = 4.1 \times 10^4$:** A decimal point is implied (assumed there) after the last 0 in 41,000. Move the decimal place four spaces to the left, creating the number 4.1. The exponent is +4.

- ✔ **$312,000,000,000 = 3.12 \times 10^{11}$:** The decimal place is moved 11 spaces to the left.

- ✔ **$0.00000031 = 3.1 \times 10^{-7}$:** The decimal place is moved seven spaces to the right this time. This is a very small number, and the exponent is negative.

- ✔ **$0.2 = 2 \times 10^{-1}$:** The decimal place is moved one space to the right.

Exploring exponential expressions

Expressing very large numbers or very small numbers exponentially makes them so much easier to deal with! Exponents are very helpful when studying situations that involve doing the same thing over and over again. These events are often called sequences or series.

Picture a cat stalking a mouse. They're about 100 centimetres apart. Every time the mouse starts nibbling at a hunk of cheese, the cat takes advantage of the mouse's distraction and creeps closer by one-tenth the distance between them. The cat wants to get about 6 centimetres away — close enough to pounce (it's a cautious cat). How far apart are they after four moves? How about after ten moves? How long will it take before the cat can pounce on the mouse?

Use these steps to stalk your own mouse (or to figure any decreasing distance):

1. **Express the incremental move as a fraction.**

   In the sample problem, that's easy because the cat creeps closer by one-tenth the distance between them.

2. **Multiply the total distance by the fraction to get the length of the move.**

   The cat and mouse are 100 centimetres apart, so you multiply 100 times $\frac{1}{10}$ to get 10 centimetres.

3. **Subtract the length of the move from the current distance.**

   100 centimetres minus 10 centimetres leaves 90 centimetres between them.

4. **Multiply the current distance by the fraction to find the distance of the second move.**

   Second move: Multiply 90 times $\frac{1}{10}$ to get 9 centimetres.

5. **Subtract the length of the move from the current distance.**

   centimetres minus 9 centimetres leaves 81 centimetres between them.

6. **Multiply the current distance by the fraction to find the distance of the third move.**

   Third move: Multiply 81 times $\frac{1}{10}$ to get 8.1 centimetres.

7. **Subtract the length of the move from the current distance.**

   81 centimetres minus 8.1 centimetres = 72.9 centimetres between them.

And so on, and so on, and so on. (Aren't you glad the cat wasn't 200 centimetres away?)

Good news: There's an easier way. Instead of finding one-tenth the distance remaining each time and subtracting, switch to finding the distance remaining between them, which is nine-tenths of the distance before that move. One-tenth plus nine-tenths equals one — the whole amount.

In each step, you multiply by $\frac{9}{10}$, the fraction of the distance left after the move times the current distance. Nine-tenths times the current distance is the new distance. Then there's just one operation to deal with each time.

1. **Find the distance left between them after the first move by multiplying the current distance by $\frac{9}{10}$.**

   $\frac{9}{10} \times 100 = 90$ centimetres between them

2. **Find the distance left between them after the second move by multiplying the current distance by $\frac{9}{10}$.**

   $\frac{9}{10} \times 90 = 81$ centimetres between them.

3. **Find the distance left between them after the third move by multiplying the current distance by $\frac{9}{10}$.**

   $\frac{9}{10} \times 81 = 72\frac{9}{10}$ centimetres between them.

4. **Find the distance left between them after the fourth move by multiplying the current distance by $\frac{9}{10}$.**

   $\frac{9}{10} \times 72\frac{9}{10} = 0.9 \times 72.9 = 65.61$ centimetres between them.

Again, as you see, this can get pretty tedious. The best way to find the answer is to use exponents. Figuring this problem using powers, or exponents, can make the computation easier. The third time is the charm for finding the distance between the cat and the mouse. Just use this formula:

Distance to pounce $= 100\left(\frac{9}{10}\right)^n$, where n is the number of moves the cat has made

Perform the operations inside the grouping symbol first.

In this formula, because the fraction $\frac{9}{10}$ is inside parentheses, apply the exponent just outside the parentheses to the fraction first. Multiply the fraction n times itself before multiplying it by 100.

After the third move, the distance between them is $100\left(\frac{9}{10}\right)^3 = 72.9$ centimetres.

After the tenth move, the distance between them is $100\left(\frac{9}{10}\right)^{10} \approx 34.87$ centimetres. This still isn't close enough to pounce.

Note: I'm using the approximately symbol ($\approx$) here because the actual answer has many more decimal places and you don't need all that information.

After the 26th move, the distance between them is $100\left(\frac{9}{10}\right)^{26} \approx 6.46$ centimetres.

It'll take one more move to be within the 6-centimetre pounce distance. Do you suppose the mouse still hasn't caught on after 26 moves? If not, then it deserves to be pounced upon.

Now, to get away from this game of cat and mouse, let me bounce to an example that uses a bouncing ball and a formula to solve it. Here goes.

> Find the total distance that a super ball travels if it always bounces back 75 per cent of the distance it fell. You dropped it from a third-floor window that's 40 metres above a nice, smooth sidewalk. Assume that the ball always falls straight down and returns straight up. (The theoretical is always easier than the practical.) Figure 6-1 shows you some of the first drops and bounces.

If you want to find out the total distance (up and down and up and down and up . . .) that a super ball travels in n bounces, if it always bounces back 75 per cent of the distance it falls, then you want to add up all the distances — *all* of them!

In Figure 6-1, I show you some of the first distances: the original 40 metres, then 75 per cent of $40 = 30$ metres, then 75 per cent of $30 = 22.5$ metres and so on. The list of numbers 40, 30, 22.5, 16.875, 12.65625, and so on are part of an *infinite geometric sequence*. A geometric sequence is formed when each term is found by multiplying the previous term by a particular number,

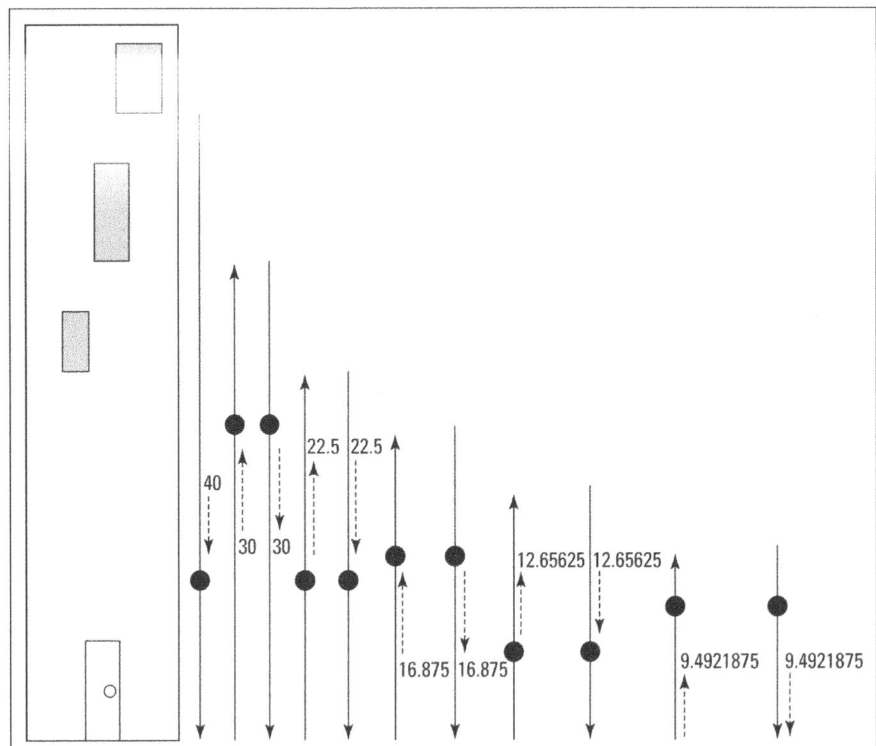

Figure 6-1: Does the bouncing ever really stop?

called the *ratio*. I don't go into all the good details, and you'll have to trust me here, but I can tell you that the sum of all the terms (infinitely many) of this type of geometric sequence is found with a rather simple formula.

The sum of the terms of an infinite geometric sequence where the ratio, r, is a number between 0 and 1, is found by dividing the first term of the sequence, a, by the difference between 1 and r:

$$\text{Sum} = \frac{40}{1-r}$$

In this bouncing-ball problem, I first add up all the downward distances. The first term is the 40, and the ratio is 0.75. So, using the formula, you get the sum of $40 + 30 + 22.5 + 16.875 + \ldots$ to be:

$$\text{Sum} = \frac{40}{1-0.75} = \frac{40}{0.25} = 160$$

That sum just gives you all the distances going downward. What about all the bounces back up? The simplest way to find all the upward distances is to just subtract 40 from the sum of the downward distances. (The number 40 is the only number not repeated.) So, subtracting $160 - 40$, you get 120. And, adding the downward distances to the upward distances, you get $160 + 120 = 280$ metres. The ball travels a total distance of 280 metres.

You may wonder what the bouncing ball has to do with exponents — the main topic of this section. The answer is that the distance that the ball travels in n bounces is found with a formula involving powers of the 75 per cent bounce return. Using the equation:

$$\text{distance} = 40 + 240\,[1 - 0.75^{n}]$$

You can find the distance after two bounces, four bounces, ten bounces, and so on. I already showed you the total number of bounces possible when n is *forever*, so anything less than *forever* will be less than 280 metres. After one bounce and before the second bounce, the total distance is 40 metres + 30 metres + 30 metres = 100 metres. Check this with the formula:

$$\text{distance} = 40 + 240\,[1 - 0.75^{1}] = 100 \text{ metres}$$

After ten bounces, the total distance is 40 metres + 30 metres + 30 metres + 22.5 metres + $\ldots$ Ugh! Use the formula!

$$\text{distance} = 40 + 240\,[1 - 0.75^{10}] \approx 266.48 \text{ metres}$$

The total will keep getting closer and closer to 280 metres. Your calculator will round the total to 280 before you even get to $n = 100$.

Okay, now you want to know where the formula for the distance came from.

The sum of the first n terms of *any* geometric sequence is found with this formula:

Sum $= \frac{a(1-r^n)}{1-r}$, where a is the first term in the sequence, r is the ratio, and n is the number of terms

With this formula, you aren't restricted to the ratio being between 0 and 1 or by n being infinitely large (all the terms). For the bouncing ball problem, I want to double each term after the 40, so I put the 40 in front and add it to twice the sum of the next n terms or bounces. The first term then becomes 30 and the ratio is still 0.75.

$$
\begin{aligned}
\text{Sum} &= 40 + 2\left[\frac{30(1-0.75^n)}{1-0.75}\right] \\
&= 40 + 2\left[\frac{30(1-0.75^n)}{0.25}\right] \\
&= 40 + 2\left[\frac{30}{0.25}(1-0.75^n)\right] \\
&= 40 + 2\left[120(1-0.75^n)\right] \\
&= 40 + 240(1-0.75^n)
\end{aligned}
$$

Voilà!

Chapter 7

Working with the Variability of Variables

In This Chapter

▶ Understanding how variables always stand in for numbers

▶ Simplifying expressions using addition, subtraction, multiplication and division

▶ Expanding expressions to ensure equal shares

▶ Mixing your positives and negatives and working with fractional powers

▶ Distributing over two, three (or more) terms

*W*hen algebra uses variables to represent numbers that can be added, subtracted, multiplied and divided, you always assume that the variables are representing *quantities* or *amounts* that can be added, subtracted, multiplied and divided. But using the representation is not quite that simple or obvious. Even when you're just adding numbers together, restrictions exist. Likewise, restrictions and rules exist when you're adding variables together or adding numbers and variables together.

Wait a minute! What *restrictions* are there for just adding *numbers* together? Why would there be any problem with that? Doesn't 1 plus 1 still equal 2? Sure, unless it's equal to 4. Seriously, imagine you're visiting the US and using quarters (25¢ piece) and dimes (10¢). Consider what happens when you add six quarters and four dimes. When you add the *numbers* together, what do you get (aside from not enough money for a gourmet cup of coffee)? Ten quarters? Ten dimes? Ten duarters? Ten quimes?

No. This is silly, of course. But it illustrates what I mean by restrictions on adding numbers together. If you want to add quarters and dimes, count the number of coins and say that you have ten *coins,* or change to the money value of each coin and say that you have $1.90. When adding quantities or amounts, you have to be sure that the amounts *can* be added. That's even more critical when you add letters because silly errors aren't as obvious. You have to be careful to add them correctly.

In this chapter, I take you through the rules and restrictions that apply.

Representing Numbers with Letters

The letters in the word *Anzac* represent 'Australian and New Zealand Army Corps', and NATO stands for North Atlantic Treaty Organization. What does this have to do with algebra? In algebra, letters don't stand for actions, places or people. The letters, or *variables*, always stand in for numbers.

Letting a variable represent a quantity, you can simplify a problem and lead to nice, neat situations because you don't have to deal with a bunch of messy words. Sometimes you do have to deal with some fairly complicated situations. But fear not! A few simple rules can help change even the most complicated situation into an easily understandable one.

Consider the task of collecting, organising and reporting on the coins collected during a charity drive. Let n represent the value of the 5¢ pieces in a roll, let t represent the value of the 10¢ pieces in a roll, and let f represent the value of the 50¢ pieces in a roll.

Notice that each of the variables n, t, and f represents a money amount — a number. Now let me show you the way in which the variables are used to describe multiple amounts of the numbers.

I'm putting you in charge of combining all the efforts of the charity-drive helpers. After collecting the money, you get this information from your helpers:

- Ann collected six rolls of 5¢ pieces, four rolls of 10¢ pieces and nine rolls of 50¢ pieces, or $6n + 4t + 9f$.

- Ben collected five rolls of 5¢ pieces, three rolls of 10¢ pieces and seven rolls of 50¢ pieces, or $5n + 3t + 7f$.

- Cal collected 15 rolls of 5¢ pieces, two rolls of 10¢ pieces and six rolls of 50¢ pieces, or $15n + 2t + 6f$.

- Don collected one roll of 5¢ pieces, three rolls of 10¢ pieces and four rolls of 50¢ pieces, or $1n + 3t + 4f$.

Now you can combine the amounts, compare the amounts or sort the amounts using shorthand notation involving the rolls of coins. (The section

'Adding and subtracting variables,' later in this chapter, shows you how to do the maths.)

When you want to add terms, because each of them has an *a*, you can add them together as long as the *a* represents the same thing in each one. One *a* can't represent the number of apples while the other *a* represents the number of aardvarks. They all have to represent the same thing in the same problem.

A variable that appears more than once in an expression or equation should always represent the same number. If the variable could represent more than one thing, the statement would be worthless — with no way to tell one meaning of the variable from another.

It's nice when the variable chosen to represent some number can start with the same letter as what it represents, such as *a* for aardvark. But this isn't necessary. A letter/name coordination is useful when a problem is composed of more than one variable, but taking careful notes and identifying variables works just as well.

Attaching factors and coefficients

One nice thing about algebra is that it conserves energy — the energy that would be needed to write multiple multiplication symbols between letters or between a number and a letter. Even having to write teeny little dots between symbols takes time, so a simpler system was devised. When a number is written in front of a variable, such as $3x$, it means that the 3 and *x* are multiplied together. The 3 and the *x* are both factors of the term $3x$. And the 3 is a *coefficient* in this case — it gets a special designation.

A number preceding a variable is a *coefficient*. For example, the number 4 is the coefficient when $a + a + a + a$ is expressed as $4a$. When several variables are multiplied together, multiplication symbols aren't needed. The term $3xyz$ means that all four factors are multiplied together.

Interpreting the operations

The symbols + and − may mean many things to you. You look for the symbols on your batteries when inserting them in a torch. You always want to see the + symbol when looking at your bank-account balance. The + and − symbols mean several things in algebra, too. The meaning of the symbol all depends on the context. For one thing, the + and − symbols always separate *terms*, which are clusters of variables and numbers connected by multiplication and division. In algebra-speak, a plus sign means *and, more, increased by, added to* and so on. For example:

- +*a*: A gift of *a* dollars has *increased* the value of my account by *a* dollars.

- 2 + *a*: Two people went through the door, and then *a more* went in.

- *a* + 20: The temperature was *a* degrees, and then it went *up* 20 degrees.

This is somewhat different from the term 2*a* in which the coefficient 2 doubles the amount of the variable *a*:

- 2*a*: Hillary lost *a* pounds, but Georgia lost 2*a*, *twice* as much.

- 2*a*: The temperature was *a* degrees, and then it *doubled* to 2*a* degrees.

The minus sign means *less, take away, decreased by, subtracted from* and so on.

- *a* − 2: There were *a* administrators, but their number was *decreased* by two.

- *a* − 1: There was one *less* than *a* alligators in the pond.

- *a* − 4: William Tell had *a* apples when he started and four *fewer* when he finished.

Even though you have to take care when letting variables represent numbers, the benefit and ease in working with variables outweighs the possible difficulties. Besides the advantage of not having to write as much, focusing on a problem that takes up less space is easier — your eyes can track better. Also, algebraic symbols are precise. The hidden meanings that written words can have don't exist in algebra. Algebra is a universal language that crosses the boundaries that language and time can present.

Doing the Maths

Addition was probably the first operation you discovered. Addition is the easiest for people of all ages to picture and relate to, and it's usually the happiest operation. 'Do you want one more cookie? How many does that make?!' Adding is a bit trickier in algebra, just because you often come to places where you *can't* add. But when you can, it's a nice process. When *can't* you add? You can't add *a* + *b* to get an *ab* in the same way that you can't add apples and bananas to get apanas.

Adding and subtracting variables

When adding like variables, instead of expressing $a + a + a + a$ the long way, you can just write $4a$, which says the same thing more efficiently because multiplication is just repeated addition. In the case of $4a$, the number represented by a is added four times. Or you can say that a is multiplied by 4.

When adding or subtracting terms that have *exactly* the same variables, combine the coefficients.

When adding $2a + 5a + 4a$ what is the result?

$$2a + 5a + 4a = (2 + 5 + 4)a = 11a$$

Why does this work? Just look at the three terms in another way:

$$2a = a + a \qquad 5a = a + a + a + a + a \qquad 4a = a + a + a + a$$

So $2a + 5a + 4a = a + a + a + a + a + a + a + a + a + a + a = 11a$.

That's a total of $11a$ variables altogether. Notice that the numbers in front — the coefficients 2, 5 and 4 — add up to 11.

When a variable has no number in front, assume that the coefficient is a 1:

$$a = 1a \qquad x = 1x$$

The following examples show you how one variable can be added to another term with the same variable or variables.

Simplify the following expression by combining like terms: $a + 3a + x + 2x$.

$$a + 3a + x + 2x =$$
$$1a + 3a + 1x + 2x =$$
$$(1 + 3)a + (1 + 2)x = 4a + 3x$$

Notice that you add terms that have the same variables because they represent the same amounts. You don't try to add the terms with different variables.

Here's another example. Simplify the following expression by combining like terms: $3x + 4y - 2x - 8y + x$.

$$3x + 4y - 2x - 8y + x =$$
$$(3 - 2 + 1)x + (4 - 8)y = 2x - 4y$$

When subtracting terms, use the rules for adding and subtracting signed numbers and apply them to the coefficients. (Check out Chapter 3 for information on working with signed numbers.)

For example, simplify the following expression by combining like terms: $5az + 4az - 2a + 6 - 3b - 2b$.

$$5az + 4az - 2a + 6 - 3b - 2b =$$
$$(5 + 4)az - 2a + (-3 - 2)b + 6 = 9az - 2a - 5b + 6$$

Notice that the 6 doesn't have a variable. It stands by itself and isn't multiplying anything. Also, a term with az is different from a term with just a, so they don't combine.

Adding and subtracting with powers

The following list of simplifications shows how addition and subtraction are performed on several terms involving variables with exponents:

- $x + x + x = 3x$
- $x^2 - 2x^2 + 3x^2 + 3x^2 = 5x^2$
- $x + 3x + 4x^2 + 5x^2 + 6x^3 = 4x + 9x^2 + 6x^3$
- $4x^4 - 3x^3 + 2x^2 + x - 1 = 4x^4 - 3x^3 + 2x^2 + x - 1$

Notice that the terms that combine *always* have exactly the same variables with exactly the same powers. (For more on powers, or exponents, refer to Chapter 6.) In the last problem, none of the powers are the same, so even though the variables are the same, you can't add the numbers in front together.

In order to add or subtract terms with the same variable, the exponents of the variable must be the same. Perform the required operations on the coefficients, leaving the variable and exponent as they are. Because x and x^2 don't represent the same amount, they can't be added together.

Simplify the following expression with powers: $3a^3 + 3a^2 + 3a + a + 2a^2 + 2a^4$.

$$3a^3 + 3a^2 + 3a + a + 2a^2 + 2a^4 =$$
$$2a^4 + 3a^3 + (3 + 2)\, a^2 + (3 + 1)a = 2a^4 + 3a^3 + 5a^2 + 4a$$

Notice that the exponents are listed in order from highest to lowest. This is a common practice to make answers easy to compare.

As another example, simplify the following expression with powers: $2m + 3m^2 + 5m^3 - 2m^2 - 3m - 1$.

$$2m + 3m^2 + 5m^3 - 2m^2 - 3m - 1 =$$
$$5m^3 + (3 - 2)\, m^2 + (-3 + 2)\, m - 1 = 5m^3 + m^2 - m - 1$$

Multiplying and Dividing Variables

Multiplying variables is in some ways easier than adding or subtracting them, just as with fractions — multiplying and dividing fractions is easier than adding or subtracting because you don't have to find common denominators. The only real caution comes when you divide variables; you need to follow some relatively strict rules to avoid dividing by 0. In this section, I give you the tips and rules.

Multiplying variables

When the variables are the same in a multiplication problem, multiplying them together 'compresses' them into a single factor, or variable. You're able to write the expression in a shorter format by using powers. But, as with addition and subtraction, you still can't combine *different* variables.

When multiplying factors containing variables, multiply the coefficients and variables as usual. If the bases are the same, you can multiply the bases by merely adding their exponents. (See more on the multiplication of exponents in Chapter 6.)

Here are some examples of multiplying several variable factors:

- ✔ $a \cdot a \cdot b \cdot c = a^2bc$: The two factors a combine with an exponent to show the number of times the factor appears in the expression.

- ✔ $2 \cdot a \cdot a \cdot a \cdot b \cdot b \cdot c = 2a^3b^2c$

- ✔ $2 \cdot a \cdot a \cdot a \cdot a \cdot 3 \cdot b \cdot b \cdot b \cdot 4 \cdot c \cdot c = 24a^4b^3c^2$: The three numbers have a product of 24. Multiplication is commutative, so you can multiply them in any order.

> ✔ $2 \cdot a^2 \cdot a^3 \cdot 3 \cdot b \cdot b \cdot b^6 \cdot 5 \cdot c \cdot c^2 \cdot c^{10} = 30a^5 b^8 c^{13}$: Add the exponents on the like factors.
>
> ✔ $(2a^2\, b^2\, c^3)(4a^3\, b^2\, c^4) = 2(4)a^{2+3}\, b^{2+2}\, c^{3+4} = 8a^5\, b^4\, c^7$
>
> ✔ $(3x^2\, yz^{-2})(4x^{-2}\, y^2\, z^4)(3xyz) = 3(4)(3)x^{2-2+1}\, y^{1+2+1}\, z^{-2+4+1} = 36x^1\, y^4\, z^3$

Dividing variables

When you want to divide one term containing variables and numbers by another, divide the numbers as if you're reducing fractions (refer to Chapter 4 for fraction reduction). But only variables that are alike can be divided.

In division of whole numbers, such as $27 \div 5$, the answers don't have to come out even. There can be a *remainder* (a value left over when one number is divided by another). But you usually don't want remainders when dividing algebraic expressions — the remainders would be new terms. So, be sure you don't leave any remainders lying around.

When dividing variables, write the problem as a fraction. Using the greatest common factor (GCF), divide the numbers and reduce. Use the rules of exponents (refer to Chapter 6) to divide variables that are the same. Dividing variables is fairly straightforward. Each variable is considered separately. The number of coefficients are reduced the same as in simple fractions.

First, let me illustrate this rule with aluminium cans. Say four friends decide to collect aluminium cans for recycling (and money — they live in South Australia). They collected $12x^3$ cans, and they're going to get y^2 cents per can. The total amount of money collected is then $12x^3 y^2$ cents. How will they divvy this up?

Divide the total amount by 4 to get the individual amount that each of the four friends will receive:

$$\frac{12x^3 y^2}{4} = 3x^3 y^2 \text{ cents each}$$

The only thing that divides here is the coefficient. If you want the number of cans each will get paid for, divide by $4y^2$ instead of just 4:

$$\frac{12x^3 y^2}{4y^2} = 3x^3 \text{ cans}$$

Why is using variables better than using just numbers in this aluminium-can story? Because if the number of cans or the value per can changes, you still have all the shares worked out. Just let the x and y change in value.

The following examples show how to divide using variables, coefficients and exponents. Use division of algebraic expressions to solve the following problems:

- ✔ If a stands for the number of apples, ten apples divided into groups of five apples each results in two groups (not two apples): $\frac{10a}{5a} = 2$.

- ✔ Ten apples divided into five groups results in two apples per group: $\frac{10a}{5} = 2a$.

- ✔ Simplify the expression $\frac{6a^2}{3a} = 2a$. Three divides 6 twice. Using the rules of exponents, $a^2 \div a = a$.

- ✔ Simplify $\frac{14x^2}{7x^4} = 2x^{-2} = \frac{2}{x^2}$. I prefer to write the answer with x in the denominator and a positive exponent rather than in the numerator with a negative exponent, but either answer is equally correct.

- ✔ And one more simplification: $\frac{6x^3y^2}{18xy^4} = \frac{x^2}{3y^2}$.

Doing it all

I cover the four main operations — addition, subtraction, multiplication and division — in the preceding sections. But many algebra problems involve more than one operation, so look at the following steps to see how to handle a combination of operations.

In this next problem, you see multiplication and addition. The order of operations still applies (refer to Chapter 5), and the rules for combining factors and terms are in force.

Simplify: $4a^2\,b^3\,(2a^3\,b^2) + 5ab^{-2}\,(2a^4\,b^7) + 5$.

1. **Rearrange the factors in each term so you can multiply the variables together separately.**

   $$4 \times 2a^2a^3b^3b^2 + 5 \times 2aa^4b^{-2}b^7 + 5$$

2. **Multiply the numbers and add the exponents of the variables that are alike.**

   $$8a^{2+3}\,b^{3+2} + 10a^{1+4}\,b^{-2+7} + 5 = 8a^5\,b^5 + 10a^5\,b^5 + 5$$

   You can see that the first two terms are alike as far as the variables they have and the exponents on those variables, which is why you can add them together.

3. Combine terms that are alike.

$$(8 + 10)a^5 b^5 + 5 = 18a^5 b^5 + 5$$

Okay. Now that you've successfully met the challenge of performing several operations on one complex example, why not try going through the steps again to perform a combination of operations on another example?

Simplify: $3m^2(2mn) - 4m^3 n^3(2n^{-2}) + 5m^2 n^3 - 6mn(mn)$.

1. **Rearrange the factors in each term so you can multiply the variables together separately.**

   $$3 \times 2m^2 mn - 4 \times 2m^3 n^3 n^{-2} + 5m^2 n^3 - 6mmnn$$

2. **Multiply the numbers and add the exponents of the variables that are alike.**

   $$6m^{2+1} n - 8m^3 n^{3-2} + 5m^2 n^3 - 6m^{1+1} n^{1+1} = 6m^3 n - 8m^3 n$$
   $$+ 5m^2 n^3 - 6m^2 n^2$$

3. **Combine the terms that are alike.**

   In this case, only the first two terms can be combined; their variables and their exponents match.

   $$(6 - 8)\, m^3 n + 5m^2 n^3 - 6m^2 n^2 = -2m^3 n + 5m^2 n^3 - 6m^2 n^2$$

The following example is your chance to strut your stuff. You've done the multiplying, so the next step is division (which is really simple subtraction). Go for it!

Simplify: $\dfrac{4x^2 y^3}{2xy} - \dfrac{15xy^5}{3y^3} + \dfrac{13x^{-2}y^{11}}{x^{-5}y^8} + \dfrac{11x^4 y^{\frac{7}{2}}}{xy^{\frac{1}{2}}}$.

1. **Divide by subtracting the exponents of the common bases.**

   Divide the known numbers. Assume that the base without an exponent has 1 for an exponent. This problem has negative exponents to deal with in both the numerator and the denominator.

   $$2x^{2-1}y^{3-1} - 5xy^{5-3} + 13x^{-2+5}y^{11-8} + 11x^{4-1}y^{\frac{7}{2}-\frac{1}{2}}$$

2. **Complete the subtraction on the exponents.**

   Note: When the negative exponent x^{-5} that was in the denominator was brought up, it became positive and was added. Fractional exponents work just like other whole-number exponents; they add and subtract just the same.

   $$2xy^2 - 5xy^2 + 13x^3 y^3 + 11x^3 y^3$$

3. **Add or subtract the terms that are exactly alike — numbers that have variables and exponents in common.**

$$(2 - 5)xy^2 + (13 + 11)x^3 y^3 = -3xy^2 + 24x^3 y^3$$

Expanding Expressions

Algebra is full of contradictory actions. First, you're asked to reduce fractions, and then you're supposed to multiply and create bigger numbers. First, you're asked to simplify or compress the maths expression, and then you're asked to spread it all out again. Make up your mind!

But rest assured that good reasons are behind doing all these seemingly contradictory processes. You carefully wrap a birthday gift so it can be unwrapped the next day. Your parents (okay, probably your dad) water and fertilise your lawn to make it grow — just so they can cut it. See, contradictions are everywhere! In this section, I take you through expanding algebraic expressions, so items can be shared.

Getting your equal share

When things are shared equitably, everyone or everything involved gets an equal share — just one of the shares — not twice as many as others get. When a child is distributing her birthday treats to classmates, it's, 'One for you, and one for you ...' Any other way is cheating! In algebra, distributing is much the same process — each gets a share.

Distributing items is an act of spreading them out equally. Algebraic distribution means to multiply each of the terms within the parentheses by another term that is outside the parentheses. Each term gets multiplied by the same amount.

To distribute a term over several other terms, multiply each of the other terms by the first. Distribution is multiplying each individual term in a grouped series of terms by a value outside of the grouping.

$$a(b + c + d + e + \ldots) = ab + ac + af + ae + \ldots$$

The addition signs could just as well be subtraction, and a is any real number: Positive, negative, integer, fraction.

A *term* is made up of variable(s) and/or number(s) joined by multiplication and/or division. Terms are separated from one another by addition or subtraction.

As an example, distribute the number 2 over the terms $4x + 3y - 6$.

1. **Multiply each term by the number(s) and/or variable(s) outside of the parentheses.**

   $2(4x + 3y - 6)$

   $2(4x) + 2(3y) - 2(6)$

2. **Perform the multiplication operation in each term.**

   $= 8x + 6y - 12$

When you distribute some factor over several terms, you don't change the value of the original expression. The answer is the same, whether you distribute first or add up what's in the parentheses first. When performing algebraic manipulations, you often have to make a judgement call as to whether to combine what's in the parentheses first or to distribute first.

Distributing first to get the answer is the better choice when the multiplication of each term gives you nicer numbers. Fractions or decimals in the parentheses are sometimes changed into nice, whole numbers when the distribution is done first. The other choice — adding up what's in the parentheses first — is preferred when the distributing gives you too many big multiplication problems. Sometimes it's easy to tell which case you have; other times, you just have to guess and try it.

Distributing first

Sometimes you can tell just by looking whether it's easier to distribute the term outside the parentheses before or after summing up the terms within the parentheses. Taking a moment to see what may be the best approach may save time in the long run.

Simplify by finding the product: $60\left(\frac{1}{2} + \frac{3}{5} - \frac{3}{4} + \frac{13}{15}\right)$.

Look at what's involved if you multiply 60 times $\frac{1}{2} + \frac{3}{5} - \frac{3}{4} + \frac{13}{15}$ only after adding the fractions first. You need to find a common denominator and then add and subtract the fractions:

$$60\left(\frac{30}{60} + \frac{36}{60} - \frac{45}{60} + \frac{52}{60}\right) =$$

$$60\left(\frac{73}{60}\right) = 73$$

Now look at the better choice, where the distribution is done first:

$$60\left(\frac{1}{2}+\frac{3}{5}-\frac{3}{4}+\frac{13}{15}\right)$$

Multiplying by 60 gets rid of all the fractions, so you don't have to find a common denominator.

$$60\left(\frac{1}{2}\right)+60\left(\frac{3}{5}\right)-60\left(\frac{3}{4}\right)+60\left(\frac{13}{15}\right)=$$

$$30 + 36 - 45 + 52 = 73$$

Do you see the advantage in this case of doing the distribution first? For an example of a situation in which doing the adding first is best, see the next section.

Adding first

Before working through a distribution problem, look at the size of the numbers. If the numbers are large, distributing one large term over other large terms within the parentheses can only make each term larger and less manageable. In the case of big numbers, it may be easier to work through any simple addition and subtraction within the parentheses before distributing the term outside the parentheses over those within.

For example, simplify: $43(160 - 159 + 433 - 432)$.

Distribute 43 first (ugh):

$$43(160 - 159 + 433 - 432) =$$
$$43(160) - 43(159) + 43(433) - 43(432) =$$
$$6{,}880 - 6{,}837 + 18{,}619 - 18{,}576 = 86$$

Now look at the better choice, where you combine first:

$$43(160 - 159 + 433 - 432) =$$
$$43(1+1) = 43(2) = 86$$

The examples in this section and the preceding section are a bit exaggerated, but I wanted to make a point. The best route to take isn't always obvious. But if you keep your eyes open for the choices available, you can save yourself some time and some work.

Distributing Signs

When a number is distributed over terms within parentheses, you multiply each term by that number. An even easier type of distribution is distributing a simple sign (no, you don't distribute a Leo or Libra — they'd object). But, what should be rather simple is often done in error. Hence, I'm devoting some time to signs.

Positive (+) and negative (–) signs are simple to distribute, but distributing a negative sign can cause errors.

Distributing positives

Distributing a positive sign makes no difference in the signs of the terms. For example, distribute positive numbers over terms:

- $+(4x + 2y - 3z + 7)$ is the same as multiplying through by +1:

$$+1(4x + 2y - 3z + 7) = +1(4x) + 1(2y) + (-3z) + 1(7)$$
$$= 4x + 2y - 3z + 7$$

- When distributing +3, the signs of the terms don't change:

$$+3(4x + 2y - 3z + 7) = +3(4x) + 3(2y) + 3(-3z) + 3(7)$$
$$= 12x + 6y - 9z + 21$$

Even when a positive number other than the number 1 is distributed, it doesn't affect the signs. The terms were changed by the multiplier of 3, but the signs of the terms in the expression stayed the same.

Distributing negatives

When distributing a negative sign, each term has a change of sign: From negative to positive or from positive to negative. For example, distributing -1 over terms in the parentheses $-(4x + 2y - 3z + 7)$ is the same as multiplying through by -1:

$$-1(4x + 2y - 3z + 7) =$$
$$-1(4x) - 1(2y) - 1(-3z) - 1(7) = -4x - 2y + 3z - 7$$

Each term was changed to a term with the opposite sign.

One mistake to avoid when you're distributing a negative sign is not distributing over *all* the terms. This is especially the case when the process is *hidden.* By hidden, I mean that a negative sign may not be in front of the whole expression, where it sticks out. It can be between terms, showing a subtraction and not being recognised for what it is. Don't let the negative signs ambush you.

For example, simplify the expression by distributing and combining like terms: $4x(x - 2) - (5x + 3)$.

Distribute the $4x$ over the x and the -2 by multiplying both terms by $4x$:

$$4x(x - 2) = 4x(x) - 4x(2)$$

Distribute the negative sign over the $5x$ and the 3 by changing the sign of each term. Be careful — you can easily make a mistake if you stop after only changing the $5x$.

$$-(5x + 3) = -(+5x) - (+3)$$

Multiply and combine the like terms:

$$4x(x) - 4x(2) - (+5x) - (+3) =$$
$$4x^2 - 8x - 5x - 3 = 4x^2 - 13x - 3$$

Reversing the roles in distributing

Distributing multiplication over an expression that has several terms added or subtracted is an extension of simply multiplying. What does this do to the value of an expression in terms of the commutative law?

Multiplication is *commutative,* which means that the order in which you multiply the terms doesn't matter: $a \times b = b \times a$.

Palindromes

The word *palindrome* comes from the Greek word *palindromos*, which means *running back again*. A palindrome is any word, sentence or even a complete poem that reads the same backward as it does forward. For example, Leigh Mercer wrote, 'A man, a plan, a canal — Panama' to honour the man responsible for building the Panama Canal. Or, how about 'Niagara, O roar again!' Words can also be palindromes: *rotator, Malayalam* (an East Indian language) and *redivider.*

Number palindromes have been of great interest to mathematicians over the years. Some perfect squares are palindromes: 121 and 14,641 for example. A palindromic date might be 1 September 1901 (1091901). Some couples choose their wedding dates by observing when a particular day is a palindrome.

You can create a palindrome by reversing the digits of almost any number and adding the reversal to the original number. For example, take 146: Reverse the digits to get 641. Add them together: 146 + 641 = 787. If you don't get a palindrome, just repeat the steps (and repeat ...) until you finally (and you will) get a palindrome.

What happens to distributing if you reverse the order? After all, both adding and subtracting is involved, too.

For example, find and compare the products: $3(x^2 + y - 7 - z)$ and $(x^2 + y - 7 - z)3$.

In the first expression, the 3 is in front:

$$3(x^2 + y - 7 - z) =$$
$$3(x^2) + 3(y) + 3(-7) + 3(-z)$$

Giving you:

$$3x^2 + 3y - 21 - 3z$$

In the second expression, the three is in back.

$$(x^2 + y - 7 - z)3 =$$
$$x^2(3) + y(3) - 7(3) - z(3)$$

Multiply and rewrite:

$$3x^2 + 3y - 21 - 3z$$

The results are exactly the same. Hurrah! A task made easier.

Mixing It Up with Numbers and Variables

Distributing variables over the terms in an algebraic expression involves multiplication rules and the rules for exponents. When different variables are multiplied together, they can be written side by side without using any multiplication symbols between them. If the same variable is multiplied as part of the distribution, the exponents are added together.

When multiplying factors with the same base, add the exponents:

$$a^x \cdot a^y = a^{x+y}$$

Let me show you a couple of distribution problems involving factors with exponents.

Distribute the a through the terms in the parentheses: $a(a^4 + 2a^2 + 3)$.

Multiply a times each term:

$$a(a^4 + 2a^2 + 3) =$$
$$a \cdot a^4 + a \cdot 2a^2 + a \cdot 3$$

Use the rules of exponents to simplify:

$$a^5 + 2a^3 + 3a$$

As another example distribute z^4 over the terms $2z^2 - 3z^{-2} + z^{-4} + 5z^{\frac{1}{3}}$.

Distribute the z^4 by multiplying it times each term:

$$z^4(2z^2 - 3z^{-2} + z^{-4} + sz^{\frac{1}{3}}) =$$
$$z^4 \cdot 2z^2 - z^4 \cdot 3z^{-2} + z^4 z^{-4} + z^4 \cdot 5z^{\frac{1}{3}}$$

Simplify by adding the exponents:

$$2z^{4+2} - 3z^{4-2} + z^{4-4} + 5z^{4+\frac{1}{3}} =$$
$$2z^6 - 3z^2 + z^0 + 5z^{\frac{13}{3}} = 2z^6 - 3z^2 + 1 + 5z^{\frac{13}{3}}$$

The exponent 0 means the value of the expression is 1. $x^0 = 1$ for any real number x except 0.

You combine exponents with different signs by using the rules for adding and subtracting signed numbers. Fractional exponents are combined after finding common denominators. Exponents that are improper fractions are left in that form.

This next example shows what happens when you have more than one variable — and how you have to use the rule of adding exponents very carefully.

Simplify the expression by distributing:
$5x^2y^3(16x^2 - 2x + 3xy + 4y^3 - 11y^5 + z - 1)$.

Multiply each term by $5x^2y^3$:

$$5x^2y^3 \cdot 16x^2 - 5x^2y^3 \cdot 2x + 5x^2y^3 \cdot 3xy + 5x^2y^3 \cdot 4y^3 - 5x^2y^3 \cdot 11y^5$$
$$+ 5x^2y^3z - 5x^2y^3$$

Complete the multiplication in each term. Add exponents where needed:

$$80x^4y^3 - 10x^3y^3 + 15x^3y^4 + 20x^2y^6 - 55x^2y^8 + 5x^2y^3z - 5x^2y^3$$

You're finished! There are no like terms that can be combined.

The next example is cluttered with negative signs.

Simplify by distributing: $-4xyzw(4 - x - y - z - w)$.

Multiply each term by $-4xyzw$:

$$-4xyzw(4) - 4xyzw(-x) - 4xyzw(-y) - 4xyzw(-z) - 4xyzw(-w)$$

Complete the multiplication in each term:

$$-16xyzw + 4x^2yzw + 4xy^2zw + 4xyz^2w + 4xyzw^2$$

Negative exponents yielding fractional answers

As the heading suggests, a base that has a negative exponent can be changed to a fraction. The base and the exponent become part of the denominator of the fraction, but the exponent loses its negative sign in the process. Then you cap it all off with a 1 in the numerator.

The formula for changing negative exponents to fractions is $a^{-n} = \frac{1}{a^n}$. (Refer to Chapter 6 for more details on negative exponents.)

In the following example, I show you how a negative exponent leads to a fractional answer.

Distribute the $5a^{-3}b^{-2}$ over each term in the parentheses:

$$5a^{-3}b^{-2}(2ab^3 - 3a^2b^2 + 4a^4b - ab) =$$

$$5a^{-3}b^{-2}(2ab^3) - (5a^{-3}b^{-2})(3a^2b^2) + (5a^{-3}b^{-2})(4a^4b) - (5a^{-3}b^{-2})(ab)$$

Multiplying the numbers and adding the exponents:

$$10a^{-3+1}b^{-2+3} - 15a^{-3+2}b^{-2+2} + 20a^{-3+4}b^{-2+1} - 5a^{-3+1}b^{-2+1}$$

The factor of b with the 0 exponent becomes 1:

$$10a^{-2}b^1 - 15a^{-1}b^0 + 20a^1b^{-1} - 5a^{-2}b^{-1}$$

This next step shows the final result without negative exponents — using the formula for changing negative exponents to fractions (see earlier in this section). Note that either answer is equally correct.

$$\frac{10b}{a^2} - \frac{15}{a} + \frac{20a}{b} - \frac{5}{a^2b}$$

Working with fractional powers

Exponents that are fractions work the same way as exponents that are integers. When multiplying factors with the same base, the exponents are added together. The only hitch is that the fractions must have the same denominator to be added. (The rules don't change just because the fractions are exponents.)

For example, distribute and simplify: $x^{\frac{1}{4}}y^{\frac{2}{3}}\left(x^{\frac{1}{2}} + x^{\frac{3}{4}}y^{\frac{1}{3}} - y^{-\frac{1}{3}}\right)$.

Multiply the factor times each term:

$$x^{\frac{1}{4}}y^{\frac{2}{3}} \cdot x^{\frac{1}{2}} + x^{\frac{1}{4}}y^{\frac{2}{3}} \cdot x^{\frac{3}{4}}y^{\frac{1}{3}} - x^{\frac{1}{4}}y^{\frac{2}{3}} \cdot y^{-\frac{1}{3}}$$

Rearrange the variables and add the exponents:

$$x^{\frac{1}{4}}x^{\frac{1}{2}}y^{\frac{2}{3}} + x^{\frac{1}{4}}x^{\frac{3}{4}}y^{\frac{2}{3}}y^{\frac{1}{3}} - x^{\frac{1}{4}}y^{\frac{2}{3}}y^{-\frac{1}{3}} =$$

$$x^{\frac{1}{4}+\frac{1}{2}}y^{\frac{2}{3}} + x^{\frac{1}{4}+\frac{3}{4}}y^{\frac{2}{3}+\frac{1}{3}} - x^{\frac{1}{4}}y^{\frac{2}{3}-\frac{1}{3}}$$

Finish up by adding the fractions:

$$x^{\frac{3}{4}}y^{\frac{2}{3}} + x^1y^1 - x^{\frac{1}{4}}y^{\frac{1}{3}}$$

Radicals can be changed to expressions with fractions as exponents, as introduced in Chapter 6. This is handy when you want to combine terms with the same bases and you have some of the bases under radicals:

- $\sqrt{x} = x^{\frac{1}{2}}$

- $\sqrt{xy} = \sqrt{x}\sqrt{y} = x^{\frac{1}{2}}y^{\frac{1}{2}}$

- $\sqrt{x^3} = \left(x^3\right)^{\frac{1}{2}} = x^{\frac{3}{2}}$

- $\sqrt[n]{a} = a^{\frac{1}{n}}$

Distribution is easier when you have radicals in the problem if you first change everything to fractional exponents. (Turn to Chapter 6 for more on exponential operations within radicals.)

Remember: The exponent rule for raising a product in parentheses to a power is to multiply each power in the parentheses by the outside power — for example: $(x^4y^3)^2 = x^8y^6$.

Simplify by distributing: $\sqrt[n]{a^m} = a^{\frac{m}{n}}$.

Change the radical notation to fractional exponents:

$$\sqrt{xy^3}\left(\sqrt{x^5 y} - \sqrt{xy^7}\right)$$

Raise the powers of the factors inside the parentheses:

$$\sqrt{xy^3}\left(\sqrt{x^5 y} - \sqrt{xy^7}\right) = \left(xy^3\right)^{\frac{1}{2}}\left[\left(x^5 y\right)^{\frac{1}{2}} - \left(xy^7\right)^{\frac{1}{2}}\right]$$

Distribute the outside term over each term within the parentheses:

$$x^{\frac{1}{2}}y^{\frac{3}{2}}\left[x^{\frac{5}{2}}y^{\frac{1}{2}} - x^{\frac{1}{2}}y^{\frac{7}{2}}\right]$$

Add the exponents of the variables:

$$x^{\frac{1}{2}}y^{\frac{3}{2}}\left(x^{\frac{5}{2}}y^{\frac{1}{2}}\right) - x^{\frac{1}{2}}y^{\frac{3}{2}}\left(x^{\frac{1}{2}}y^{\frac{7}{2}}\right) = x^{\frac{6}{2}}y^{\frac{4}{2}} - x^{\frac{2}{2}}y^{\frac{10}{2}}$$

Simplify the fractional exponents:

$$x^3y^2 - x^1y^5$$

Binomials and Trinomials: Distributing More Than One Term

The preceding sections in this chapter describe how to distribute one term over several others. This section shows you how to distribute a *binomial* (a polynomial with two terms). You also discover how to distribute polynomials with three or more terms.

The word *polynomial* comes from *poly* meaning 'many' and *nomen* meaning 'name' or 'designation'. A polynomial is an algebraic expression with one or more terms in it. For example, a polynomial with one term is a *monomial*; a polynomial with two terms is a *binomial*. If you have three terms, it's a *trinomial*.

Distributing binomials

Distributing two terms (a *binomial*) over several terms amounts to just applying the distribution process twice. The following steps tell you how to distribute a binomial over some polynomial:

1. **Break the binomial into its two terms.**

2. **Distribute each term of the binomial over the other factor.**

3. **Do the distributions you've created.**

4. **Simplify and combine any like terms.**

For example, multiply using distribution: $(x^2 + 1)(y - 2)$.

1. **Break the binomial into its two terms.**

   In this case, $(x^2 + 1)(y - 2)$, break the first binomial into its two terms, x^2 and 1.

2. **Distribute each term over the other factor.**

   Multiply the first term, x^2, times the second binomial, and multiply the second term, 1, times the second binomial.

   $$x^2(y - 2) + 1(y - 2)$$

3. **Do the two distributions.**

   $$x^2(y - 2) + 1(y - 2) = x^2y - 2x^2 + y - 2$$

4. Simplify and combine any like terms.

In this case, nothing can be combined; none of the terms are alike.

Now that you have the idea, try walking through a polynomial distribution that has variables in all the terms.

Multiply using distribution: $(a^2 + 2b)(4a^2 + 3ab - 2ab^2 - b^2)$.

Break the binomial into its two terms and multiply those terms times the second factor:

$$a^2(4a^2 + 3ab - 2ab^2 - b^2) + 2b(4a^2 + 3ab - 2ab^2 - b^2)$$

Doing the two distributions:

$$a^2(4a^2) + a^2(3ab) - a^2(2ab^2) - a^2(b^2) + 2b(4a^2) + 2b(3ab) - 2b(2ab^2) - 2b(b^2)$$

Multiply and simplify if possible:

$$4a^4 + 3a^3b - 2a^3b^2 - a^2b^2 + 8a^2b + 6ab^2 - 4ab^3 - 2b^3$$

Distributing trinomials

A *trinomial* (a polynomial with three terms) can be distributed over another expression. Each term in the first factor is distributed separately over the second factor, and then the entire expression is simplified, combining anything that can be combined. This process can be extended to multiplying with any size polynomial. In this section, I show you trinomials and leave the extension to you when you run into anything larger.

The following problem introduces you to working through the distribution of trinomials.

Multiply by distributing the first factor over the second: $(x + y + 2)$ $(x^2 - 2xy + y + 1)$.

Distribute each term of the trinomial by multiplying them by the second factor:

$$x(x^2 - 2xy + y + 1) + y(x^2 - 2xy + y + 1) + 2(x^2 - 2xy + y + 1)$$

Do the three distributions:

$$x^3 - 2x^2y + xy + x + x^2y - 2xy^2 + y^2 + y + 2x^2 - 4xy + 2y + 2$$

Simplify:

$$x^3 - x^2y + 2x^2 + x - 2xy^2 + y^2 - 3xy + 3y + 2$$

Multiplying a polynomial by another polynomial

This is where I establish a rule that can cover just about any product of any number of terms. You can use this general method for four, five or even more terms.

When distributing a polynomial (many terms) over any number of other terms, multiply each term in the first factor times each of the terms in the second factor. When the distribution is done, combine anything that goes together to simplify.

$$(a + b + c + d + \ldots)(z + y + x + w + \ldots) =$$
$$az + ay + ax + aw + \ldots + bz + by + bx + bw + \ldots + cz + cy + cx + cw + \ldots$$

For example, multiply the two trinomials by distributing: $(x^2 + x + 2)$ $(3x^2 - x + 1)$.

Separate the terms in the first factor from one another. Multiply each term in the first factor times the second factor:

$$(x^2 + x + 2)(3x^2 - x + 1) =$$
$$x^2(3x^2 - x + 1) + x(3x^2 - x + 1) + 2(3x^2 - x + 1)$$

Distribute and do the multiplication:

$$3x^4 - x^3 + x^2 + 3x^3 - x^2 + x + 6x^2 - 2x + 2$$

Combine like terms if possible:

$$3x^4 + 2x^3 + 6x^2 - x + 2$$

Making Special Distributions

Several distribution shortcuts can make life easier. Distributing binomials over other terms is not difficult, but you can save time if you recognise situations where you can apply a shortcut. If you don't notice that a special shortcut could have been used, don't worry about your oversight. But you may end up kicking yourself afterwards for not taking advantage of the easier process.

Recognising the perfectly squared binomial

When the same binomial is multiplied by itself — when each of the first two terms is distributed over the second and same terms — then the resulting trinomial contains the squares of the two terms and twice their product: $(a + b)^2 = (a + b)(a + b) = a^2 + 2ab + b^2$.

For example, square the binomial using the special rule: $(x + 3)^2 = (x + 3)(x + 3)$.

The result of the following operation is the sum of the squares of x and 3 along with twice their product:

- ✔ The square of x is x^2.

- ✔ The square of 3 is 9.

- ✔ Twice the product of x and 3 is $2(x \cdot 3) = 6x$.

So:

$$(x + 3)^2 =$$
$$(x + 3)(x + 3) = x^2 + 6x + 9$$

Notice that the preferred order of the terms was used: decreasing powers of x.

As another example, square the binomial. See $\left(4x - \frac{1}{3}\right)^2$.

- ✔ The square of $4x$ is $(4x)^2 = 16x^2$.

- ✔ The square of $\left(-\frac{1}{3}\right)^2 = \frac{1}{9}$. Note that this square is positive.

- ✔ Twice the product of $4x$ and $-\frac{1}{3}$ is $2 \times 4x \times \left(-\frac{1}{3}\right) = -\frac{8}{3}x$ or $\left(-2\frac{2}{3}x\right)$.

So:

$$\left(4x - \frac{1}{3}\right)^2 = 16x^2 - \frac{8}{3}x + \frac{1}{9}$$

This rule works very nicely when the terms have high powers.

Here's another example. Square the binomial using the special rule: $(2x^4 + y^3)^2$.

- ✔ The square of $2x^4$ is $(2x^4)^2 = 4x^8$.
- ✔ The square of y^3 is $(y^3)^2 = y^6$.
- ✔ Twice the product of $2x^4$ and y^3 is $2(2x^4 \cdot y^3) = 4x^4y^3$.

So:

$$(2x^4 + y^3)^2 = 4x^8 + 4x^4y^3 + y^6$$

This special method is also handy when your binomial contains another binomial!

Square the expression using the special rule: $[x + (a + b)]^2$.

- ✔ The square of x is x^2.
- ✔ The square of the binomial $(a + b)$ is $(a + b)^2 = a^2 + 2ab + b^2$.
- ✔ Then you find that twice the product of x and $(a + b)$ is $2x(a + b)$.

Putting all the results together, you have:

$$[x + (a + b)]^2 =$$
$$x^2 + 2x(a + b) + a^2 + 2ab + b^2 = x^2 + 2xa + 2xb + a^2 + 2ab + b^2$$

Spotting the sum and difference of the same two terms

Just one little — which can be *big* — difference arises between the multiplications in this section and the ones in the preceding section. The difference is that there's a sign change between the first and second binomials. Instead of multiplying exactly the same binomial times itself, in this section you do a switcheroo and see two different signs separating the terms. The same two terms are always used — it's just that the sign between them changes.

The sum of any two terms multiplied by the difference of the same two terms is easy to spot and even easier to work out.

The sum of any two terms multiplied by their difference equals the difference of the squares of these same two terms. For any real numbers a and b: $(a + b)(a - b) = a^2 - b^2$.

Notice that the middle term just disappears because a term and its opposite are always in the middle. You can see that here, where the terms in the first binomial are distributed over the second:

$$(a + b)(a - b) =$$
$$a(a - b) + b(a - b)$$

Multiplying and simplifying:

$$a^2 - ab + ab - b^2 = a^2 - b^2$$

The rule always works, so you can use the shortcut to do these special distributions.

For example, find the product using the special rule: $(x - 4)(x + 4)$.

- ✔ The first term squared is x^2.
- ✔ The second term will always be negative and a perfect square like the first term: $(-4)(+4) = -16$.

So:

$$(x - 4)(x + 4) = x^2 - 16$$

Here's another example. Find the product using the special rule: $(ab - 5)(ab + 5)$. This problem has a slightly more complicated variable term.

- ✔ The square of $ab = (ab)^2 = a^2b^2$.
- ✔ The opposite of the square of $5 = -25$.

So:

$$(ab - 5)(ab + 5) = a^2b^2 - 25$$

Find the product using the special rule: $[5 + (a - b)][5 - (a - b)]$. In this problem, the second term is a binomial.

> ↗ The square of $5 = 25$.
>
> ↗ The opposite of the square of $(a - b) = -(a - b)^2$.

Square the binomial and distribute the negative sign:

$$-(a - b)^2 - (a^2 - 2ab + b^2) =$$
$$-a^2 + 2ab - b^2$$

So:

$$[5 + (a - b)][5 - (a - b)] =$$
$$25 - a^2 + 2ab - b^2$$

Chapter 8

Smaller is Better: Factoring Down

In This Chapter

▶ Preparing for perplexing prime numbers

▶ Bringing big numbers down to size

▶ Investigating composite numbers with prime factorisations

▶ Using variables versus numbers and finding the greatest common factor

▶ Getting terms joined together by grouping

*P*rime numbers (whole numbers evenly divisible only by themselves and one) have been the subject of discussions between mathematicians and non-mathematicians for centuries. Prime numbers and their mysteries have intrigued philosophers, engineers and astronomers. These folks and others have discovered plenty of information about prime numbers, but many unproven conjectures remain. Prime numbers play an important role in coding (encrypting passwords and protecting information).

Probably the biggest mystery is determining what prime number will be discovered next. Computers have aided the search for a comprehensive list of prime numbers, but because numbers go on forever without end, and because no-one has yet found a pattern or method for listing prime numbers, the question involving the *next big one* remains.

You may believe in the bigger-is-better philosophy, which can apply to spending money, cookies or television screens, but it doesn't really work for algebra. For the most part, the opposite is true in algebra: Smaller numbers are easier and more comfortable to deal with than larger numbers.

In this chapter, I take you through some of the basics relating to prime numbers, and reducing down to the prime factor. You then discover how to get to those smaller-is-better terms. You find the basics of factoring and how factoring is related to division. The factoring patterns you see here carry over somewhat in more complicated expressions.

Beginning with the Basics

Prime numbers are important in algebra because they help you work with the smallest-possible numbers. Big numbers are often unwieldy and can produce more computation errors when you perform operations and solve equations. So, reducing fractions to their lowest terms and factoring expressions to make problems more manageable are basic and very desirable tasks.

A *prime number* is a whole number larger than the number 1 that can be divided evenly only by itself and 1.

The first and smallest prime number is the number 2. It's the only *even* prime number. All primes after 2 are odd because all even numbers can be divided evenly by 1, themselves and 2. So even numbers greater than 2 don't fit the definition of a prime number.

Here are the first 46 prime numbers:

2	3	5	7	11	13	17	19
23	29	31	37	41	43	47	53
59	61	67	71	73	79	83	89
97	101	103	107	109	113	127	131
137	139	149	151	157	163	167	173
179	181	191	193	197	199		

Why isn't the number 1 prime?

By tradition and definition, the number 1 is not prime. The definition of a prime number is that it can be divided evenly only by itself and 1. In this case, there would be a double hit, because 1 is itself.

Many theorems and conjectures involving primes don't work if 1 is included. Mathematicians around the time of Pythagoras sometimes even excluded the number 2 from the list of primes because they didn't consider 1 or 2 to be *true numbers* — they were just generators of all other even and odd numbers. Sometimes it seems that mathematical rules are a bit arbitrary. But in this case, it just makes everything else work better if 1 isn't a prime.

When you already recognise that a number is prime, you don't waste time trying to find things to divide into it when you're reducing a fraction or factoring an expression. There are so many primes that you can't memorise or recognise them all, but just knowing or memorising the primes smaller than 100 is a big help, and memorising the first 46 (all the primes smaller than 200) would be a bonus.

Composing Composite Numbers

Prime numbers are interesting to think about, but they can also be a dead end in terms of factoring algebraic expressions or reducing fractions. The opposite of prime numbers, *composite numbers*, can be broken down into factorable, reducible pieces. In this section, you see how every composite number is the product of prime numbers, in a process known as *prime factorisation*. Every number's prime factorisation is unique.

The *prime factorisation* of a number is the unique product of prime numbers that results in the given number. A prime number's prime factorisation consists of just that prime number, by itself.

Here are some examples of prime factorisation:

- $6 = 2 \cdot 3$
- $12 = 2 \cdot 2 \cdot 3 = 2^2 \cdot 3$
- $16 = 2 \cdot 2 \cdot 2 \cdot 2 = 2^4$
- $250 = 2 \cdot 5 \cdot 5 \cdot 5 = 2 \cdot 5^3$
- $510{,}510 = 2 \cdot 3 \cdot 5 \cdot 7 \cdot 11 \cdot 13 \cdot 17$
- $42{,}059 = 137 \cdot 307$

Okay, so that last one is a doozy. Finding that prime factorisation without a calculator, a computer or list of primes is difficult.

The factors of some numbers aren't always obvious, but I do have some techniques to help you write prime factorisations, so check out the next section.

Writing Prime Factorisations

Writing the prime factorisation of a composite number is one way to be absolutely sure you've left no stone unturned when reducing fractions or factoring algebraic expressions. These factorisations show you the one and only way a number can be factored. Two favourite ways of creating prime factorisations are upside-down division and trees.

Dividing while standing on your head

A slick way of writing out prime factorisations is to do an upside-down division. You put a *prime factor* (a prime number that evenly divides the number you're working on) on the outside left and the result or *quotient* (the number of times it divides evenly) underneath. You divide the quotient (the number underneath) by another prime number and keep doing this until the bottom number is a prime. Then you can stop. The order you do the divisions in doesn't matter. You get the same result or list of prime factors no matter what order you use. So, if you like to get all the even factors out first, just divide by 2 until you can't any longer.

Here's an example of finding the prime factorisation of 120 using upside-down division:

$$2\lfloor 120$$
$$2\lfloor 60$$
$$2\lfloor 30$$
$$3\lfloor 15$$
$$5$$

Starting with the only even prime, I divided by 2, first, and got 60. Because 60 isn't prime, I divided by 2 again and got 30. Then I divided by 2 again and got 15. The number 15 is divisible by 3, and the result of the division is 5. Because the number 5 is prime, I stopped dividing and used the results to write the prime factorisation.

Looking at the numbers going down the left side of the work and the number at the bottom, you see that they act the same as the divisors in a division problem — only, in this case, they're all prime numbers. Although many composite numbers could have played the role of divisor for the number 120, the numbers for the prime factorisation of 120 must be prime-number divisors.

When using this process, you usually do all the 2s first, then all the 3s, then all the 5s, and so on to make the prime factorisation process easier, but you can do this in any order: $120 = 2 \cdot 2 \cdot 2 \cdot 3 \cdot 5 = 2^3 \cdot 3 \cdot 5$. In the next example, start with 13 because it seems obvious that it's a factor. The rest are all in a mixed-up order.

Here's another prime factorisation example, this time finding the prime factorisation of 13,000:

$$13\overline{)13,000}$$
$$5\overline{)1000}$$
$$2\overline{)200}$$
$$2\overline{)100}$$
$$5\overline{)50}$$
$$2\overline{)10}$$
$$5$$

So $13,000 = 13 \cdot 5 \cdot 2 \cdot 2 \cdot 5 \cdot 2 \cdot 5 = 2^3 \cdot 5^3 \cdot 13$.

Getting to the root of primes with a tree

Another popular method for finding prime factorisations is to use a tree. Think of the number you start with as being the trunk of the tree and the prime factors as being at the ends of the roots.

To use the tree method, you write down your number and find two factors with a product is that number. Then you find factors for the two factors, and factors for the factors of the factors and so on. You're finished when the lowest part of any root system is a prime number. Then you collect all those prime numbers for the factorisation.

Figure 8-1 shows an example of finding the prime factorisation of 6,350,400 using a factor tree.

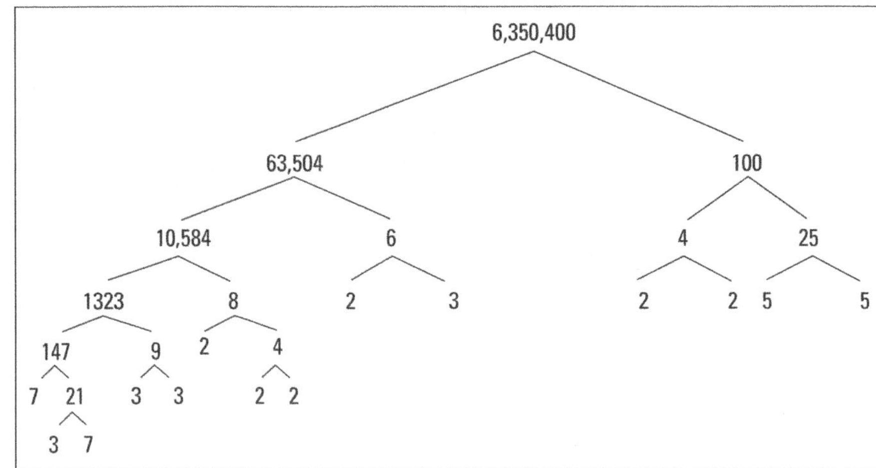

Figure 8-1:
Finding the prime factors using a tree.

Now you collect all the prime numbers at the ends of the roots. I see the prime factors as: 7, 3, 7, 3, 3, 2, 2, 2, 2, 3, 2, 2, 5, 5. Putting them in order, I get:

$$6,350,400 = 2 \cdot 2 \cdot 2 \cdot 2 \cdot 2 \cdot 2 \cdot 3 \cdot 3 \cdot 3 \cdot 3 \cdot 5 \cdot 5 \cdot 7 \cdot 7$$

$$= 2^6 \cdot 3^4 \cdot 5^2 \cdot 7^2$$

You may not have created a tree the same way I did. Everyone sees different multiples and factors and has his or her favourites as far as dividing. I like to stick to numbers I can divide in my head. But you may be a calculator person. The great thing is that every way works and gives you the same final answer.

Wrapping your head around the rules of divisibility

The techniques for finding prime factorisations work just fine, as long as you have a good head start on what divides a number evenly. You probably already know the rules for dividing by 2 or 5 or 10. But many other numbers have very helpful rules or gimmicks for just looking at the number and seeing whether it's divisible by a particular factor. In Table 8-1, I give you many of the more commonly used rules of divisibility. Some are easier to use than others. Notice that I don't have the numbers in order; I prefer to group the numbers by the types of rules used.

Table 8-1	Rules of Divisibility
Number	**Rule**
2	The number ends in 0, 2, 4, 6 or 8.
5	The number ends in 0 or 5.
10	The number ends in 0.
4	The last two digits form a number divisible by 4.
8	The last three digits form a number divisible by 8.
3	The sum of the digits is a number divisible by 3.
9	The sum of the digits is a number divisible by 9.
11	The difference between the sums of the alternating digits is divisible by 11.
6	The number is divisible by both 2 and 3 (use both rules).
12	The number is divisible by both 3 and 4 (use both rules).

Here's an example of using the rules of divisibility to determine what divides 360 evenly:

- The number 360 ends in 0, so it's divisible by 2, 5 and 10.

- The last two digits of 360 form the number 60, which is divisible by 4, so the whole number is divisible by 4.

- The last three digits of 360 (okay, so it's all of them) form a number divisible by 8, so the whole number is divisible by 8.

- The sum of the digits in 360 is 9, so the number is divisible by both 3 and 9.

- The difference between the sums of the alternating digits is 3, so the number is not divisible by 11. (To get that difference, I added the $3 + 0$ to get 3 and the 6 had nothing to add to it. The difference between 6 and 3 is 3.)

- The number 360 is divisible by 2 and 3, both, so it's divisible by 6.

- The number 360 is divisible by 3 and 4, both, so it's divisible by 12.

Here's another example, this time with the number 1,056:

- ✔ The number 1,056 ends in 6, so it's divisible by 2.
- ✔ The last two digits of 1,056 form the number 56, which is divisible by 4, so the whole number is divisible by 4.
- ✔ The last three digits of 1,056 form the number 56 (the 0 in front is ignored), which is divisible by 8, so the whole number is divisible by 8.
- ✔ The sum of the digits is 12, which is divisible by 3, so the whole number is divisible by 3.
- ✔ The difference between the sums of the alternating digits is 0, which is divisible by 11, so the whole number is divisible by 11.
- ✔ The number is divisible by 2 and 3, both, so it's divisible by 6.
- ✔ The number is divisible by 3 and 4, both, so it's divisible by 12.

Rules also apply for 7 and 13 and other multiples of numbers, but the additional rules are seldom used, so I don't go into them here.

Getting Down to the Prime Factor

Doing the actual factoring in algebra is easier when you can recognise which numbers are composite and which are prime. If you know in which category a number belongs, you know what to do with it. When reducing fractions or factoring out many-termed expressions, you look for what the numbers have in common. If a number is prime, you stop looking. Now, try putting all this knowledge to work!

Taking primes into account

Prime factorisations are useful when you reduce fractions. Sure, you can do repeated reductions — first divide the numerator and denominator by 5 and then divide them both by 3 and so on. But a much more efficient use of your time is to write the prime factorisations of the numerator and denominator and then have an easy task of finding the common factors all at once.

For example, reduce the fraction $\frac{120}{165}$ by following these steps:

1. Find the prime factorisation of the numerator.

120 is $2^3 \cdot 3 \cdot 5$.

2. Find the prime factorisation of the denominator.

165 is $3 \cdot 5 \cdot 11$.

3. Next, write the fraction with the prime factorisations in it.

$$\frac{120}{165} = \frac{2^3 \cdot 3 \cdot 5}{3 \cdot 5 \cdot 11}$$

4. Cross out the factors the numerator shares with the denominator to see what's left — the reduced form.

$$\frac{120}{165} = \frac{2^3 \cdot 3 \cdot 5}{3 \cdot 5 \cdot 11} = \frac{2^3 \cdot \cancel{3} \cdot \cancel{5}}{\cancel{3} \cdot \cancel{5} \cdot 11} = \frac{2^3}{11} = \frac{8}{11}$$

Now, try reducing the fraction $\frac{100}{243}$:

1. Find the prime factorisation of the numerator.

100 is $2^2 \cdot 5^2$.

2. Find the prime factorisation of the denominator.

243 is 3^5.

3. Write the fraction with the prime factorisations.

$$\frac{100}{243} = \frac{2^2 \cdot 5^2}{3^5}$$

Look at the prime factorisations. You can see that the numerator and denominator have absolutely nothing in common. The fraction can't be reduced. The two numbers are *relatively prime*. The beauty of using the prime factorisation is that you can be sure that the fraction's reduction possibilities are exhausted — you haven't missed anything. You can leave the fraction in this factored form or go back to the simpler $\frac{100}{243}$. It depends on your preference.

Next, I add some variables to the mix.

Reduce the fraction $\frac{48x^3y^2z}{84xy^2z^3}$:

1. Find the prime factorisation of the numerator.

$48x^3y^2z = 2^4 \cdot 3 \cdot x^3y^2z$.

2. Find the prime factorisation of the denominator.

$84xy^2z^3 = 2^2 \cdot 3 \cdot 7 \cdot xy^2z^3$.

3. Write the fraction with the prime factorisation.

$$\frac{48x^3y^2z}{84xy^2z^3} = \frac{2^4 \cdot 3 \cdot x^3y^2z}{2^2 \cdot 3 \cdot 7 \cdot xy^2z^3}$$

4. Cross out the factors in common.

$$\frac{2^4 \cdot 3 \cdot x^3 \cdot y^2 \cdot z}{2^2 \cdot 3 \cdot 7 \cdot x \cdot y^2 \cdot z^3} = \frac{2^{4^2} \cdot 3 \cdot x^{3^2} \cdot y^2 \cdot z}{2^2 \cdot 3 \cdot 7 \cdot x \cdot y^2 \cdot z^{3^2}} = \frac{2^2 \cdot x^2}{7 \cdot z^2} = \frac{4x^2}{7z^2}$$

By writing the prime factorisations, you can be certain that you haven't missed any factors that the numerator and denominator may have in common.

As another example, reduce the fraction $\frac{5,400,000,000,000,000,000}{711,000,000,000,000,000,000}$:

1. Find the prime factorisation of the numerator.

Take advantage of all the zeros by writing the number using a variation on scientific notation.

$$5,400,000,000,000,000,000 = 54 \times 10^{17}$$
$$= 2 \cdot 3^3 \times 10^{17}$$

2. Find the prime factorisation of the denominator, using scientific notation again.

$$711,000,000,000,000,000,000,000 = 711 \times 10^{21}$$
$$= 3^2 \times 79 \times 10^{21}$$

3. Write the fraction with the prime factorisation.

$$\frac{2 \times 3^3 \times 10^{17}}{3^2 \times 79 \times 10^{21}}$$

4. Cross out the factors in common.

$$\frac{2 \times 3^3 \times 10^{17}}{3^2 \times 79 \times 10^{21}} = \frac{2 \times 3^{3^1} \times 10^{17}}{3^2 \times 79 \times 10^{21^4}} = \frac{2 \times 3}{79 \times 10^4}$$

The 2 in the numerator and the 10 in the denominator have a factor of 2 in common:

$$\frac{2\times3}{79\times2\times5\times10^3} = \frac{\cancel{2}\times3}{79\times\cancel{2}\times5\times10^3} = \frac{3}{79\times5\times10^3} = \frac{3}{395,000}$$

Pulling out factors and leaving the rest

Pulling out common factors from lists of terms or the sums or differences of a bunch of terms is done for a good reason. It's a common task when you're simplifying expressions and solving equations. The common factor that makes the biggest difference in these problems is the greatest common factor (GCF). When you recognise the GCF and factor it out, it does the most good.

The *greatest common factor* is the largest-possible number that evenly divides each term of an expression containing two or more terms (or evenly divides the numerator and denominator of a fraction).

In any factoring discussion, the GCF, the most common and easiest factoring method, always comes up first. And it's helpful to know about the GCF when solving equations. In an expression with two or more terms, finding the greatest common factor can make the expression more understandable and manageable.

When simplifying expressions, the best-case scenario is to recognise and pull out the GCF from a list of terms. Sometimes, though, the GCF may not be so recognisable. It may have some strange factors, such as 7, 13 or 23. It isn't the end of the world if you don't recognise one of these numbers as being a multiplier; it's just nicer if you do.

The three terms in the expression $12x^2y^4 + 16xy^3 - 20x^3y^2$ have common factors. What is the GCF? These steps help you find it.

1. **Determine any common numerical factors.**

2. **Determine any common variable factors.**

3. **Write the prime factorisations of each term.**

4. **Find the GCF.**

5. **Divide each term by the GCF.**

6. **Write the result as the product of the GCF and the results of the division.**

For example, here's how to find the GCF of $12x^2y^4 + 16xy^3 - 20x^3y^2$ and write the factorisation:

1. **Determine any common numerical factors.**

   Each term has a coefficient that is divisible by a power of 2, which is $2^2 = 4$.

2. **Determine any common variable factors.**

   Each term has x and y factors.

3. **Write the prime factorisations of each term.**

   $$12x^2y^4 = 2^2 \cdot 3 \cdot x^2y^4$$

   $$16xy^3 = 2^4 \cdot xy^3$$

   $$-20x^3y^2 = -2^2 \cdot 5 \cdot x^3y^2$$

4. **Find the GCF.**

   The GCF is the product of all the factors that all three terms have in common. The GCF contains the *lowest* power of each variable and number that occurs in any of the terms. Each variable in the sample problem has a factor of 2. If the lowest power of 2 that shows in any of the factors is 2^2, then 2^2 is part of the GCF.

   Each factor has a power of x. If the lowest power of x that shows up in any of the factors is 1, then x^1 is part of the GCF.

   Each factor has a power of y. If the lowest power of y that shows in any of the factors is 2, then y^2 is part of the GCF.

   The GCF of $12x^2 y^4 + 16xy^3 - 20x^3 y^2$ is $2^2 xy^2 = 4xy^2$.

5. **Divide each term by the GCF.**

   The respective terms are divided as shown:

   - $$\frac{12x^2y^4}{4xy^2} = 3xy^2$$

   - $$\frac{16xy^3}{4xy^2} = 4y$$

   - $$\frac{-20x^3y^2}{4xy^2} = -5x^2$$

Notice that the three different results of the division have nothing in common. Each of the first two results has a y and the first and third both have an x, but nothing is shared by all the results. This is the best factoring situation, which is what you want.

6. **Write the result as the product of the GCF and the results of the division.**

Rewriting the original expression with the GCF factored out and in parentheses: $12x^2 y^4 + 16xy^3 - 20x^3 y^2 = 4xy^2 (3xy^2 + 4y - 5x^2)$.

In the next examples, I show you the shortened version of these steps.

Find the GCF and write the factorisation of $40a^5x + 80a^5y - 120a^5z = 40a^5 (x + 2y - 3z)$. The factorisations of the terms are: $2^35a^5 x$, 2^45a^5y, and $- 2^33 \cdot 5a^5z$. Each term has a factor of $2^3 \cdot 5$, and a^5, so the GCF is $40a^5$, and you can write the expression as the product of the GCF and the results of dividing each term by the GCF.

As another example, find the GCF and write the factorisation of $18x^2y + 25z^3 + 49z^2$. Even though none of these terms is prime, the three terms have nothing in common — nothing that *all three* share. The following prime factorisations demonstrate:

- $18x^2y = 2 \cdot 3^2x^2y$

- $25z^3 = 5^2 z^3$

- $49z^2 = 7^2 z^2$

The last two terms do have a factor of z in common, but the first term doesn't. This expression is said to be *prime* because it can't be factored.

Getting to First Base with Factoring

As you've probably gathered, factoring is another way of saying, 'Rewrite this so everything is all multiplied together.' You usually start out with two or more terms and have to determine how to rewrite them so they're all multiplied together in some way or another. And, oh yes, the two expressions have to be equal! Why all this fuss? You rewrite expressions as products — keeping the new results equivalent to the old — so that you can perform operations on the results. Fractions reduce more easily, equations solve more easily, and answers are observed more easily when you can factor.

Reviewing the terms and rules

You'll understand factoring better if you have a firm handle on what the terms used to talk about factoring mean:

- ✔ **Term:** A group of number(s) and/or variable(s) connected to one another by multiplication or division and separated from other terms by addition or subtraction.

- ✔ **Factor:** Any of the values involved in a multiplication problem that, when multiplied together, produce a result.

- ✔ **Coefficient:** A number that multiplies a variable and tells how many of the variable.

- ✔ **Constant:** A number or variable that never changes in value.

- ✔ **Relatively prime:** Terms that have no factors in common. If the only factor that numbers share in common is 1, they're considered *relatively prime.*

Here is an illustration for all the terms I just gave: In the expression $5xy + 4z - 6$, you see three *terms.* In the first term, $5xy$, three *factors* are all multiplied together. The 5 is usually referred to as the *coefficient.* The second term has two factors, 4 and z, and the third term contains just a *constant.* The first and second terms are *relatively prime* because they have no factors in common. (The number 4 is not prime, but it's *relatively prime* to 5.)

Factoring out numbers

Factoring is the opposite of distributing; it's 'undistributing' (refer to Chapter 7 for more on distribution). When performing distribution, you multiply a series of terms by a common multiplier. Now, by factoring, you seek to find what a series of terms has in common and then take it away, dividing the common factor or multiplier out from each term. Think of each term as a numerator of a fraction, and you're finding the same denominator for each. By factoring out, the common factor is put outside parentheses or brackets and all the results of the divisions are left inside.

An expression can be written as the product of the largest number that divides all the terms evenly times the results of the divisions: $ab + ac + ad = a(b + c + d)$.

Writing factoring as division

In the trinomial $16a - 8b + 40c^2$, 2 is a common factor. But 4 is also a common factor, and 8 is a common factor. Here are the divisions of the terms by 2, 4 and 8:

$$\frac{16a}{2} - \frac{8b}{2} + \frac{40c^2}{2} = \frac{\overset{8}{\cancel{16}}a}{\cancel{2}} - \frac{\overset{4}{\cancel{8}}b}{\cancel{2}} + \frac{\overset{20}{\cancel{40}}c^2}{\cancel{2}} = 8a - 4b + 20c^2$$

$$\frac{16a}{4} - \frac{8b}{4} + \frac{40c^2}{4} = \frac{\overset{4}{\cancel{16}}a}{\cancel{4}} - \frac{\overset{2}{\cancel{8}}b}{\cancel{4}} + \frac{\overset{10}{\cancel{40}}c^2}{\cancel{4}} = 4a - 2b + 10c^2$$

$$\frac{16a}{8} - \frac{8b}{8} + \frac{40c^2}{8} = \frac{\overset{2}{\cancel{16}}a}{\cancel{8}} - \frac{\overset{1}{\cancel{8}}b}{\cancel{8}} + \frac{\overset{5}{\cancel{40}}c^2}{\cancel{8}} = 2a - b + 5c^2$$

You see that the final result, in each case, does not contain a fraction. For a number to be a *factor*, it must divide all the terms evenly. To show the results of factoring, you write the factor outside parentheses and the results of the division inside:

$$16a - 8b + 40c^2 = 2(8a - 4b + 20c^2)$$
$$16a - 8b + 40c^2 = 4(4a - 2b + 10c^2)$$
$$16a - 8b + 40c^2 = 8(2a - b + 5c^2)$$

Outlining the factoring method

The absolutely *proper* way to factor an expression is to write the prime factorisation of each of the numbers and look for the greatest common factor (GCF — refer to the section 'Getting Down to the Prime Factor', earlier in this chapter). What's really more practical and quicker in the end is to look for the biggest factor that *you can easily recognise*. Factor it out and then see if the numbers in the parentheses need to be factored again. Repeat the division until the terms in the parentheses are relatively prime.

For example, here's how to use the repeated-division method to factor the expression $450x + 540y - 486z + 216$. You see that the coefficient of each term is even, so divide each term by 2:

$$450x + 540y - 486z + 216 = 2(225x + 270y - 243z + 108)$$

The numbers in the parentheses are a mixture of odd and even, so you can't divide by 2 again. The numbers in the parentheses are all divisible by 3, but an even better choice is available: You may have noticed that the digits in the numbers in all the terms add up to 9. That's the rule for divisibility by 9, so 9 can divide each term evenly. (Refer to the earlier section 'Wrapping your head around the rules of divisibility' for more on this.) Thus,

$$2(225x + 270y - 243z + 108) = 2[9(25x + 30y - 27z + 12)]$$

Factoring in the real world

You usually use factoring when you need to reduce fractions or solve a quadratic equation. But a type of factoring comes to the rescue in several real-life situations.

For example, Stephanie wants to organise her collection of CDs (yes, some are getting a bit old, but she loves them). She wants to put some in her room, some in the lounge room, and some by the computer. She has 18 rap CDs, 24 rock CDs, 30 hip-hop CDs and 42 pop CDs. How can she divide the CDs so she has a nice balance in each location?

Writing the numbers of CDs as a sum, Stephanie has $18 + 24 + 30 + 42$ total. Each of the numbers is divisible by 2, 3 and 6, so the factorisations could be any of the following:

$$2(9 + 12 + 15 + 21)$$

$$3(6 + 8 + 10 + 14)$$

$$6(3 + 4 + 5 + 7)$$

The last factorisation shows you four relatively prime numbers within the parentheses. But Stephanie has only three locations to put her CDs in, so she'll go with 6 rap CDs, 8 rock CDs, 10 hip-hop CDs, and 14 pop CDs in each place.

Now multiply the 2 and 9 together to get

$$450x + 540y - 486z + 216 = 18(25x + 30y - 27z + 12)$$

You could have divided 18 into each term in the first place, but not many people know the multiplication table of 18. (It's a stretch even for me.) What about the coefficients of the numbers in the parentheses? None is a prime number. And several have factors in common. But no single factor divides *all* the coefficients equally. The four coefficients are *relatively prime*, so you're finished with the factoring.

Factoring out variables

Variables represent values; variables with exponents represent the powers of those same values. For that reason, variables as well as numbers can be factored out of the terms in an expression, and in this section you can find out how.

When factoring out powers of a variable, the smallest power that appears in any one term is the most that can be factored out. For example, in an expression such as $a^4b + a^3c + a^2d + a^3e^4$, the smallest power of a that appears in any term is the second power, a^2. So you can factor out a^2

from all the terms because a^2 is the greatest common factor. You can't factor anything else out of each term: $a^4b + a^3c + a^2d + a^3e^4 = a^2(a^2b + a^1c + d + a^1e^4)$.

When performing algebraic operations or solving equations, always take the time to check your work. Sometimes the check is no more than just seeing if the answer makes sense. In the case of factoring expressions, a good visual check is to multiply the factor through all the terms in the parentheses to see if you get what you started with before factoring. To perform checks on your factoring:

- ✔ Multiply through (distribute) your answer in your head to be sure that the factored form is equivalent to the original form.

- ✔ Scan the terms in parentheses to make sure that they don't share the same variable.

For example, perform the quick checks on the following factored expression:

$$x^2y^3 + x^3y^2z^4 + x^4yz = x^2y(y^2 + x^1y^1z^4 + x^2z)\ x^2y(y^2 + xyz^4 + x^2z)$$

Does your answer multiply out to become what you started with? Multiply in your head:

$$x^2y \cdot y^2 = x^2y^3 \text{ Check!}$$

$$x^2y \cdot xyz^4 = x^3y^2z^4 \text{ Check!}$$

$$x^2y \cdot x^2z = x^4yz \text{ Check!}$$

Those are the three terms in the original problem.

Now, for the second part of the quick check: Look at what's in the parentheses of your answer. The first two terms have y and the second two have x and z, but no variable occurs in all three terms. The terms in the parentheses are relatively prime. Check!

Unlocking combinations of numbers and variables

The real test of the factoring process is combining numbers and variables, finding the GCF, and factoring successfully. Sometimes you may miss a factor or two, but a second sweep through can be done and is nothing to be ashamed of when doing algebra problems. If you do your factoring in more than one step, it really doesn't matter in what order you pull out the factors. You can do numbers first or variables first. It'll come out the same.

For example, factor $12x^2y^3z + 18x^3y^2z^2 - 24xy^4z^3$.

Each term has a coefficient that's divisible by 2, 3 and 6. You select 6 as the largest of those common factors.

Each term has a factor of x. The powers on x are 2, 3 and 1. You have to select the *smallest* exponent when looking for the GCF, so the common factor is just x.

Each term has a factor of y. The exponents are 3, 2 and 4. The smallest exponent is 2, so the common factor is y^2.

Each term has a factor of z, and the exponents are 1, 2 and 3. The number 1 is smallest, so you can pull out a z from each term.

Put all the factors together, and you get that the GCF is $6xy^2z$. So:

$$12x^2y^3z + 18x^3y^2z^2 - 24xy^4z^3 = 6xy^2z(2x^1y^1 + 3x^2z^1 - 4y^2z^2)$$

Doing a quick check, you multiply through by the GCF in your head to be sure that the products match the original expression. You then do a sweep to be sure that a common factor isn't among the terms within the parentheses. Remember that you don't need to write the 1 if a variable is to the power of 1.

Here's another example. Factor $100a^4b - 200a^3b^2 + 300a^2b^2 - 400$.

The greatest common factor of the coefficients is 100. Even though the powers of a and b are present in the first three terms, none of them occurs in the last term. So you're out of luck finding any more factors. Doing the factorisation:

$$100a^4b - 200a^3b^2 + 300a^2b^2 - 400 = 100(a^4b - 2a^3b^2 + 3a^2b^2 - 4)$$

Now factor $26mn^3 - 25x^2y + 21a^4b^4mnxy$.

Even though each of the numbers is *composite* (each can be divided by values other than themselves), the three have no factors in common. The expression cannot be factored. It's considered prime.

Factor $484x^3y^2 + 132x^2y^3 - 88x^4y^5$.

In this example, even if you don't divide through by the GCF the first time, all is not lost. A second run takes care of the problem. Often, doing the factorisations in two steps is easier because the numbers you're dividing through by each time are smaller, and you can do the work in your head.

Assume that you determined that the GCF of the expression in this example is $4x^2y$. Then $484x^3y^2 + 132x^2y^3 - 88x^4y^5 = 4x^2y(121x^1y^1 + 33y^2 - 22x^2y^4)$.

Looking at the expression in the parentheses, you can see that each of the numbers is divisible by 11 and that a y is in every term. The terms in the parentheses have a GCF of $11y$.

$$4x^2y[121x^1y^1 + 33y^2 - 22x^2y^4] =$$
$$4x^2y[11y(11x + 3y^1 - 2x^2y^3)] =$$
$$(4x^2y)(11y)(11x + 3y^1 - 2x^2y^3) = 44x^2y^2(11x + 3y^1 - 2x^2y^3)$$

You can do this factorisation all at the same time, using the GCF $44x^2y^2$, but not everyone recognises the multiples of 44. Also, the factorisation could have been done in two or more steps in a different order with different factors each time. The result always comes out the same in the end.

For one more example, factor $-4ab - 8a^2b - 12ab^2$.

Each term in the expression is negative; dividing out the negative from all the terms in the parentheses makes them positive.

$$-4ab - 8a^2b - 12ab^2 = -4ab(1 + 2a^1 + 3b^1)$$

When factoring out a negative factor, be sure to change the signs of each of the terms.

Changing factoring into a division problem

You may be a whiz at dividing terms in your head, but sometimes even the mightiest find it easier to write down the terms to be factored and the common factor as a series of division problems. Yes, even I sometimes resort to reducing fractions to make the computations easier and improve my success rate.

For example, factor $480x^4y^8z^6 - 320x^6y^4z^4 - 640x^8y^5z^3$.

First, identify the GCF of the coefficients. They all end in 0, so each is divisible by 10.

Then, looking at 48, 32 and 64 (dropping the 0 at the end), you see that the numbers are all divisible by 16.

Putting the 10 and 16 together, you see a GCF of 160.

Checking out the powers of x, y and z, you see that you can divide each term by $x^4y^4z^3$.

Now, write each term in the numerator of a fraction with the greatest common factor in the denominator:

$$\frac{480x^4y^8z^6}{160x^4y^4z^3} - \frac{320x^6y^4z^4}{160x^4y^4z^3} - \frac{640x^8y^5z^3}{160x^4y^4z^3}$$

Reducing the fractions, you get

$$\frac{\overset{3}{\cancel{480}}\,x^4\,y^{8^4}\,z^{6^3}}{\cancel{160}\,x^4\,y^4\,z^3} - \frac{\overset{2}{\cancel{320}}\,x^{6^2}\,y^4\,z^{4^1}}{\cancel{160}\,x^4\,y^4\,z^3} - \frac{\overset{4}{\cancel{640}}\,x^{8^4}\,y^{5^1}\,z^3}{\cancel{160}\,x^4\,y^4\,z^3} = \frac{3y^4z^3}{1} - \frac{2x^2z}{1} - \frac{4x^4y}{1}$$

$$= 3y^4z^3 - 2x^2z - 4x^4y$$

Notice that each term has two of the three variables, but no variable appears in all three terms. The coefficients are relatively prime, so no common factor can be pulled out. So,

$$480x^4y^8z^6 - 320x^6y^4z^4 - 640x^8y^5z^3 = 160x^4y^4z^3(3y^4z^3 - 2x^2z - 4x^4y)$$

Grouping Terms

Groups are formed when people have something in common with one another. Put 20 people on an island, leave them there for a few days, and chances are good that the 20 people will form groups as they seek out those they can relate to in some way. Television producers have capitalised on this phenomenon by creating contests on these islands and introducing all sorts of conflict and drama.

The same general process can be done in factoring (but without the drama). The rules are a bit stricter when factoring by grouping than even the island social situation, but the principle is the same. I show you those algebraic principles in this section.

When using grouping to factor, follow these steps:

1. **Divide the terms into groups of an equal number of terms in each.**

2. **Look for a GCF in each group of terms and do the factorisation.**

3. **Rewrite the expression as products of the GCF of each term and a factor in parentheses.**

4. **Look for a GCF of the new terms.**

   If the new terms don't have a GCF, try a different arrangement of the terms in the divisions.

5. **Factor out the new GCF.**

For example, factor $4xy + 4xb + ay + ab$. You see that two terms have a 4 in common. Some terms have a y in common. And some terms have a, b and x mixed in there, too. But all four terms do not have a single variable or number in common. They can be *grouped*, though, into two parts that can be factored independently:

1. **Divide the terms into groups of two terms in each.**

   Group the first two terms together and then the last two.

   $(4xy + 4xb) + (ay + ab)$

2. **Look for a GCF in each group of terms and factor.**

   $4xy + 4xb = 4x(y + b)$

   $ay + ab = a(y + b)$

3. **Rewrite the expression.**

   $4x(y + b) + a(y + b)$

4. **Look for a GCF of the new terms.**

   The new GCF is $(y + b)$.

5. **Factor out the new GCF.**

   $(y + b)(4x + a)$

The original expression containing four terms is now a single term.

This grouping business doesn't really help, though, unless the results of the two separate factorisations then share something. Looking at the preceding series of steps, in Step 3 each of the factored groups had $(y + b)$ in it. When this happens, the $(y + b)$ can be factored out of the newly formed terms:

$$4x(y + b) + a(y + b) = (y + b)(4x + a)$$

This is the factored form. If you multiply this through (distribute), you get the four terms that you started with.

Again, the following example has nothing that all the terms share in common. But, if you group the first two and the last two, you can factor those pairs.

For example, factor $ax^2y - 3a + 9x^2y - 27$.

1. **Divide the terms into equal groups of two terms in each.**

   $$(ax^2y - 3a) + (9x^2y - 27)$$

2. **Look for a GCF in each group of terms and factor.**

   $$(ax^2y - 3a) = a(x^2y - 3)$$
   $$(9x^2y - 27) = 9(x^2y - 3)$$

3. **Rewrite the expression.**

   $$a(x^2y - 3) + 9(x^2y - 3)$$

4. **Look for a GCF of the new terms.**

   The GCF is $x^2y - 3$.

5. **Factor out the new GCF.**

   $$a(x^2y - 3) + 9(x^2y - 3) = (x^2y - 3)(a + 9)$$

What happens if the terms aren't in this order? How do you know what order to write them in? Do you get a different answer? Well, scramble the terms and write the problem as $ax^2y + 9x^2y - 27 - 3a$ and see what you have.

The first two terms have a GCF of x^2y. The second two terms have a GCF of -3. Grouping and factoring gives you $x^2y(a + 9) - 3(9 + a)$.

The expressions in the parentheses don't look exactly alike, but addition is commutative — you can add in either order and get the same result. You can reverse the 9 and the a in the last factor so that it looks the same as the first:

$$x^2y\,(a + 9) - 3(a + 9)$$

Now, you can factor the $(a + 9)$ out of each term to finish the problem:

$$(a + 9)(x^2y - 3)$$

The two factors in this answer are reversed from the first way you did the problem, but multiplication is also commutative.

In this last example, note that the two pairs of terms can be grouped and factored.

Factor $4ab^2 - 8ac^2 + 5x^2b - 10x^2c$.

Grouping and factoring:

$$(4ab^2 - 8ac^2) + (5x^2b - 10x^2c) = 4a(b^2 - 2c^2) + 5x^2(b - 2c)$$

The expressions in the parentheses look similar, but they aren't the same. Changing the order won't help in this case. You now have two terms, but they don't have a common factor. This expression is as simple as it can be. In other words, it's prime (in the algebraic sense).

So far, the examples I've shown you all contained four terms, which grouped into two groups of two. What about six or eight terms? Can you use grouping? The answer is yes.

Factor $2x^2y^2 + 6x^2y + 2x^2 - 3y^2 - 9y - 3$.

Here are your choices:

- ✔ Group the first three terms together, factoring out $2x^2$; group the second three terms together, factoring out -3.
- ✔ Group the first and fourth terms together, factoring out y^2; group the second and fifth terms, factoring out $3y$; and group the third and sixth terms, just showing a multiplication of 1.

Using the first choice:

$$2x^2y^2 + 6x^2y + 2x^2 - 3y^2 - 9y - 3 = 2x^2(y^2 + 3y + 1) - 3(y^2 + 3y + 1)$$

The common factor of the two terms is then factored out.

$$(y^2 + 3y + 1)(2x^2 - 3)$$

Now, using the second method, I first have to rearrange the terms:

$$2x^2y^2 + 6x^2y + 2x^2 - 3y^2 - 9y - 3 =$$
$$(2x^2y^2 - 3y^2) + (6x^2y - 9y) + (2x^2 - 3) =$$
$$y^2(2x^2 - 3) + 3y(2x^2 - 3) + 1(2x^2 - 3)$$

You see that the three terms now all have a common factor of $(2x^2 - 3)$, which can be factored out.

$$(2x^2 - 3)(y^2 + 3y + 1)$$

The order of the two factors is different from what you get using the other method, but multiplication is commutative, so they're equivalent.

Chapter 9

Going for the Second Degree with Quadratics

In This Chapter

▶ Getting squared away with quadratic expressions

▶ Finding out how to FOIL without thwarting

▶ Stepping through the unFOIL process

▶ Getting organised to factor a quadratic

▶ Performing multiple factorisations and recognising when you're done

Quadratic (second-degree) expressions — such as $3x^2 - 12$ or $-16t^2 + 32t + 11$ — are studied extensively in algebra because they have so many applications in calculus and physics and other disciplines. These are expressions because they're made up of two or more terms with plus (+) or minus (−) signs between them. If equal signs were included, they would be equations. The good news is that they're manageable. The bad news — well, there is none! Second-degree expressions are so darned nice to work with!

Quadratics have a particular variable raised to the second degree. A quadratic expression can have one or more terms, and not all the terms must have a squared variable, but at least one of the terms needs to have that exponent of 2. Also, a quadratic expression can't have any power greater than 2 on the designated variable. The highest power in an expression determines its name.

Some quadratics may have one variable in them, such as $2x^2 - 3x + 1$. Others may have two or more variables, such as $\pi r^2 + 2\pi rh$. These expressions all have their place in mathematics and science. In this chapter, you see how they work for you.

The Standard Quadratic Expression

The quadratic, or second-degree, expression in *x* has the *x* variable that is squared, and no *x* terms with powers higher than 2. The coefficient on the squared variable is not equal to 0. The standard quadratic form is $ax^2 + bx + c$.

You may notice that the following examples of quadratic expressions all have a variable raised to the second degree:

$$4x^2 + 3x - 2 \qquad a^2 + 11 \qquad 6y^2 - 5y$$

Quadratics are usually written in terms of a variable represented by an *x*, *y*, *z* or *w*. The letters at the end of the alphabet are used more frequently for the variable, while those at the beginning of the alphabet are usually used for a number or constant. This isn't always the case, but it's the standard convention.

In the standard quadratic expression, the *a* — the coefficient of the variable raised to the second power — can't be 0. If *a* were allowed to be 0, the x^2 would be multiplied by 0, and it wouldn't be a quadratic expression anymore. The variables *b* or *c* can be 0, but *a* can't.

Quadratics don't necessarily have all positive terms either. The standard form, $ax^2 + bx + c$, is written with all positives for convenience. But if *a*, *b* or *c* represents a negative number, that term would be negative. The terms are usually written with the second-degree term first, the first-degree term next, and the number last. Another mathematical convention has to do with the order of the terms in a quadratic expression. If you find more than one variable, decide which variable makes it a quadratic expression (look for the variable that's squared) and write the expression in terms of that variable. This means, after you find the variable that's squared, write the rest of the expression in decreasing powers of that variable.

For example, rewrite $aby + cdy^2 + ef$ using the standard convention involving order. This can be a second-degree expression in *y*.

Written in the standard form for quadratics, $ax^2 + bx + c$, where the second-degree term comes first, it looks like $(cd)y^2 + (ab)y + ef$. The parentheses aren't necessary around the *cd* or the *ab* and they don't change anything, but they're used sometimes for emphasis. The parentheses just make seeing the different parts easier.

In the next example, you get to make choices.

Rewrite $a^2bx + cdx^2 + aef$ using the standard convention involving order. This can be a second-degree expression in terms of either a or x.

Writing as a second degree in a:

$$(bx)a^2 + (ef)a + cdx^2$$

Even though a second-degree factor of x is in the last term, that term is thought of as a constant, a value that doesn't change, rather than a variable if the expression is a to the second degree. Now, changing roles, the second-degree expression in x:

$$(cd)x^2 + (a^2b)x + aef$$

Reining in Big and Tiny Numbers

Some perfectly good quadratic expressions are just too awkward to handle. Some of these can be made better by factoring. Some others are just going to be uncooperative — you're stuck with them. In this section, I go back to my favourite standby for simplifying: Finding a greatest common factor (GCF). If the terms in the quadratic have something in common, that can be factored out, leaving an expression more reasonable to deal with.

For example, rewrite the following quadratics by factoring out GCFs:

- ✔ **$800x^2 + 40{,}000x - 100{,}000$:** This quadratic expression can be made more usable by factoring out the common factor and arranging the result in a nice, organised expression. It has large numbers, but each number can be evenly divided by 800 — a common factor:

  $$800x^2 + 40{,}000x - 100{,}000 = 800(x^2 + 50x - 125)$$

- ✔ **$a^2x^2 + a^2c^2 + a^2b^2x$:** These terms have four different variables with powers of 2. Only the x, though, appears in a term with a power of 1. So, you may choose to write this as a quadratic in x and factor out some of the other variables. Rewrite the expression in decreasing powers of x:

  $$a^2x^2 + a^2b^2x + a^2c^2$$

  Find the GCF, which is a^2, and factor it out:

  $$a^2(x^2 + b^2x + c^2)$$

> ✔ **$0.00000008y^2 + 0.000000004y + 0.000000016$:** This last expression consists of powers of y and multipliers that are very small. Find the GCF, which is 0.000000004, and factor it out:
>
> $$000000008y^2 + 0.000000004y + 0.000000016 = 0.000000004(20y^2 + y + 4)$$

FOILing

What is FOIL?

a) A brainchild of Mr Reynolds Aluminium.

b) An expression of dismay: 'Rats! FOILed again!'

c) An acronym for first, outer, inner and last.

Choice C is my final answer. FOIL is an acronym that cropped up somewhere between my high school years and my teaching years. It was sort of 'under the counter' at first — respected mathematicians didn't want to use it, because it seemed to introduce an unnecessary and specialised process to the work with binomials. But it has caught on and is now accepted, published, and used extensively in the algebra classroom. FOIL is easy to remember and apply.

This chapter is on factoring quadratic expressions, but first you need to find out how to multiply two binomials together using FOIL. Chapter 7 shows you how to multiply two binomials together by distributing. This chapter gives you an alternate method.

FOILing basics

Many quadratic expressions, such as $6x^2 + 7x - 3$, are the result of multiplying two *binomials* (two terms separated by addition or subtraction), so you can undo the multiplication by factoring them:

$$6x^2 + 7x - 3 = (2x + 3)(3x - 1)$$

The right side is the *factored* form. But how can you tell that the left side of that equation is equal to the right side just by looking at it? It's not like searching for a GCF, when you look for something in common. A nice way to do the multiplication is using FOIL.

What does FOIL stand for? Each of the letters refers to two different terms in the multiplication — one from each of two binomials — multiplied together in a certain order. The steps don't *have* to be done in this order, but they usually are. Otherwise, the acronym would be something like OFIL (which would be awful).

The following list describes what each letter in the FOIL acronym stands for:

- **F** stands for the *first* term in each binomial: $(3a + 6)(2a − 1)$
- **O** stands for the two *outer* terms — those farthest to the left and right: $(3a + 6)(2a − 1)$
- **I** stands for the *inner* terms in the middle: $(3a + 6)(2a − 1)$
- **L** stands for the *last* term in each binomial: $(3a + 6)(2a − 1)$

Each binomial has a left term and a right term. But the two terms have other names also (just as someone named Michael may be 'Mike' to one person and 'son' to another). The other names for the terms in the binomials refer to their positions with respect to the whole picture. The two terms not in the middle are the outer terms. The two terms in the middle are the inner terms. Use this as an example: $(a + b)(c + d)$. The terms a and c are *first*; the terms b and d are *last* in each binomial. The terms a and d are *outer*; the terms b and c are inner in the big picture. As you can see, each term has two names. In the problem $(2x + 3)(3x − 1)$ the term $2x$ is called *first* one time and *outer* another time. That's okay.

Figure 9-1 gives you a visual on how this is done.

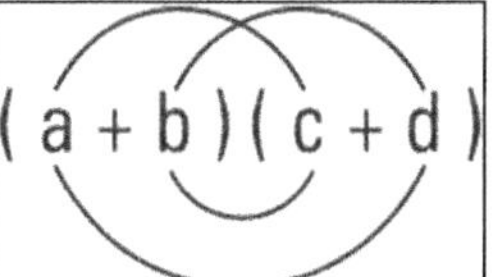

Figure 9-1: The FOIL happy face.

Algebra-speak

Just as some phrases, such as a *hill of beans,* lack a verb, an algebraic *expression,* such as $6x^2 + 11$, lacks an equal sign. Neither this phrase nor the algebraic expression makes any assertions.

On the other hand, a *statement* or an algebraic *equation* makes an assertion. A statement must contain a verb, which is similar to the equal sign or inequality symbol in an algebraic equation or inequality. For example, you may say that $6 + 3$ *is* 9. A statement, such as, 'The car is worth $15,000,' can be true or false depending on what car you're referring to. Mathematical statements — such as $6x − 1y = 11$ or $16y + 7 < 80$ — make a claim. Whether the claims are true depends on what the x and y are.

FOILed again, and again

The following steps demonstrate how to use FOIL on the problem of multiplying two binomials together: $(4x - 7)(5x + 3)$.

1. **Multiply the first term of each binomial together.**

   $(4x)(5x) = 20x^2$

2. **Multiply the outer terms together.**

   $(4x)(+3) = +12x$

3. **Multiply the inner terms together.**

   $(-7)(5x) = -35x$

4. **Multiply the last term of each expression together.**

   $(-7)(+3) = -21$

5. **List the four results of FOIL in order.**

   $20x^2 + 12x - 35x - 21$

6. **Combine the like terms.**

   $20x^2 - 23x - 21$

Distributing the two terms in the first binomial over the second produces the same result, but in the case of binomials, using FOIL is easier. (For more on distributing, refer to Chapter 7.)

See how the FOIL numbered steps work on a couple of negative terms in the following example.

Use FOIL to perform the multiplication: $(x - 3)(2x - 9)$.

1. **Multiply the first terms.**

   $(x)(2x) = 2x^2$

2. **Multiply the outer terms.**

   $(x)(-9) = -9x$

3. **Multiply the inner terms.**

   $(-3)(2x) = -6x$

4. **Multiply the last terms.**

   $(-3)(-9) = 27$

5. **List the four results of FOIL in order.**

   $2x^2 - 9x - 6x + 27$

6. **Combine the like terms.**

   $2x^2 - 15x + 27$

The following example is a bit more complicated to do, but FOIL makes it much easier. The tasks are broken down into smaller, simpler tasks, and then the results are combined for the final result.

Use FOIL to perform the multiplication: $[x + (y - 4)][3x + (2y + 1)]$.

1. **Multiply the first terms.**

   $(x)(3x) = 3x^2$

2. **Multiply the outer terms.**

   $(x)(2y + 1) = 2xy + x$

3. **Multiply the inner terms.**

   $(y - 4)(3x) = 3xy - 12x$

4. **Multiply the last terms.**

   The last terms are two binomials, too. You FOIL these binomials when you finish this series of FOIL steps.

   $(y - 4)(2y + 1)$

5. **List the four results of FOIL in order.**

   $3x^2 + 2xy + x + 3xy - 12x + (y - 4)(2y + 1)$

6. **Combine like terms.**

   $3x^2 + 5xy - 11x + (y - 4)(2y + 1)$

Now to finish the product of the two at the end: $(y - 4)(2y + 1)$. You can FOIL them:

1. **Multiply the first terms.**

   $(y)(2y) = 2y^2$

2. **Multiply the outer terms.**

   $(y)(1) = y$

3. **Multiply the inner terms.**

   $(-4)(2y) = -8y$

   4. Multiply the last terms.

   $$(-4)(1) = -4$$

   5. Write the results in order.

   $$2y^2 + y - 8y - 4$$

   6. Combine like terms.

   $$2y^2 - 7y - 4$$

Now, replace the two binomials multiplied together with this new result, and you can rewrite the entire problem:

$$3x^2 + 5xy - 11x + 2y^2 - 7y - 4$$

This may seem complicated, but using FOIL is easier than doing all the distributing.

Applying FOIL to a special product

Do you remember the rule for multiplying the sum of any two terms by their difference? If your answer is 'Yes', skip this section, give yourself a pat on the back and move to the head of the class. If your answer is 'No', just give yourself a pat on the back and keep reading.

The sum of any binomial multiplied by the difference of the same two terms (see the operation that follows) is an easy operation because the middle terms cancel each other out — they both have the same absolute value, except one is positive and the other is negative.

$$(a + b)(a - b) = a(a - b) + b(a - b) = a^2 - ab + ab - b^2 = a^2 - b^2$$

The following operation multiplies the sum and difference of the same two values. In Chapter 7, I show you how the middle terms cancel each other out or disappear. This is even more evident with FOIL.

For example, use FOIL to perform the multiplication: $(5x - 3)(5x + 3)$.

   1. Multiply the first terms.

   $$(5x)(5x) = (5x)^2 = 25x^2$$

   2. Multiply the outer terms.

   $$(5x)(3) = 15x$$

   3. Multiply the inner terms.

   $$(-3)(5x) = -15x$$

4. Multiply the last terms.

$$(-3)(+3) = -9$$

5. Write the results in order.

$$25x^2 + 15x - 15x - 9$$

6. Combine like terms.

The products, $15x$ and $-15x$, are opposites of each other. The first and last products are all that's left.

$$25x^2 - 9$$

Take a look at the multiplication problems that follow. Are they good examples of the sum and difference of binomials and FOIL?

$$(3x + 2)(3x - 2) = 9x^2 - 4$$

$$(2z - m)(2z + m) = 4z^2 - m^2$$

$$(m^2 - n^2)(m^2 + n^2) = m^4 - n^4$$

UnFOILing

When you look at an expression such as $2x^2 - 5x - 12$, you may think that figuring out how to factor this into the product of two binomials is an awful chore. And you may wonder whether it can even be factored that way. Let me assure you that these problems are really quite easy. Think of them as puzzles — not quite as challenging as Sudoku.

The nice thing in solving this particular puzzle is that a system makes unFOILing simple. You go through the system, and it helps you find what the answer is or even helps you determine if there isn't an answer. This can't be said about all factoring problems, but it is true of quadratics in the form $ax^2 + bx + c$. That's why quadratics are so nice to work with in algebra.

Unwrapping the FOILing package

The key to unFOILing these factoring problems is being organised:

✔ Be sure you have an expression in the form $ax^2 + bx + c$.

✔ Be sure the terms are written in the order of decreasing powers.

✔ If needed, review the lists of prime numbers and perfect squares.

✔ Follow the steps.

Follow these steps to factor the quadratic $ax^2 + bx + c$, using unFOIL:

1. **Determine all the ways you can multiply two numbers to get a.**

   Every number can be written as at least one product, even if it's only the number times 1. So assume that two numbers, e and f, have a product equal to a. These are the two numbers you want for this problem.

2. **Determine all the ways you can multiply two numbers together to get c.**

   If the value of c is negative, ignore the negative sign for the moment. Concentrate on what factors result in the absolute value of c.

   Now assume that two numbers, g and h, with a product equal to c. Use these two numbers for this problem.

3. **Now look at the sign of c and your lists from Steps 1 and 2.**

   - *If c is positive,* find a value from your Step 1 list and another from your Step 2 list such that the sum of their product and the product of the two numbers they're paired with in those steps results in b.

     Find $e \cdot g$ and $f \cdot h$, such that $e \cdot g + f \cdot h = b$.

   - *If c is negative,* find a value from your Step 1 list and another from your Step 2 list such that the difference of their product and the product of two numbers they're paired with from those steps results in b.

     Find $e \cdot g$ and $f \cdot h$, such that $e \cdot g - f \cdot h = b$.

4. **Arrange your choices as binomials.**

   The e and f have to be in the first positions in the binomials, and the g and h have to be in the last positions. They have to be arranged so the multiplications in Step 3 have the correct outer and inner products.

   $(e\ h)\ (f\ g)$

5. **Place the signs appropriately.**

   The signs are both positive if c is positive and b is positive.

   The signs are both negative if c is positive and b is negative.

   One sign is positive and one sign is negative if c is negative; the choice depends on whether b is positive or negative and how you arranged the factors.

For example, factor the quadratic $2x^2 - 5x - 12$ using unFOIL:

1. **Determine all the ways you can multiply two numbers to get a.**

   In this problem, a is 2, which is prime, so the only way to multiply and get 2 is 2×1.

2. **Determine all the ways you can multiply two numbers to get c.**

   In this problem, c is –12. Ignore the negative sign right now. The negative becomes important in the next step. Just concentrate on what multiplies together to give you 12.

   You can use three ways to multiply two numbers together to get 12: 12×1, 6×2 or 4×3.

3. **Look at the sign of c and your lists from Steps 1 and 2.**

   Since c is negative, you find a value from Step 1 and another from your Step 2 list such that the difference of their product and the product of the other numbers in the pairs results in b.

   Use the 2×1 from Step 1 and the 4×3 from Step 2. Multiply the 1 from Step 1 times the 3 from Step 2 and then multiply the 2 from Step 1 times the 4 from Step 2.

   $$(1)(3) = 3 \text{ and } (2)(4) = 8$$

   The two products are 3 and 8, with a difference of 5.

4. **Arrange the choices in binomials.**

   The following arrangement multiplies the $(1x)(2x)$ to get the $2x^2$ needed for the first product. Likewise, the 4 and 3 multiply to give you 12. The outer product is $3x$ and the inner product is $8x$, giving you the difference of $5x$.

   $$(1x\ 4)(2x\ 3)$$

5. **Place the signs to give the desired results.**

   You want the $5x$ to be negative, so you need the $8x$ product to be negative. The following arrangement accomplishes this:

   $$(1x - 4)(2x + 3) = 2x^2 + 3x - 8x - 12 = 2x^2 - 5x - 12$$

The next example offers many combinations of numbers to choose from.

Factor the quadratic: $24x^2 - 34x - 45$:

1. **Determine all the ways you can multiply two numbers to get 24.**

   $$24 = 24 \times 1 = 12 \times 2 = 8 \times 3 = 6 \times 4$$

2. **Determine all the ways you can multiply two numbers to get 45.**

   $45 = 45 \times 1 = 15 \times 3 = 9 \times 5$

3. **Look at the sign of *c*. The last term is −45, so you want a difference of products.**

   Use the 6×4 and the 9×5. The product of 4 and 5 is 20. The product of 6 and 9 is 54. The difference of these products is 34.

4. **Arrange your choices as binomials so the results are those you want.**

   $(4x\ 9)(6x\ 5)$

5. **Place the signs to give the desired results.**

   $(4x - 9)(6x + 5) = 24x^2 + 20x - 54x - 45 = 24x^2 - 34x - 45$

The combinations you want may not just leap out at you. But having a list of all the possibilities helps heaps. You can start systematically trying out the different combinations. For example, take the 24×1 and try it with all three sets of numbers that give you *c*: 45×1, 15×3, 9×5. If none of those work, try the 12×2 with all the sets of numbers that give you *c*. Continue until you've systematically gone through all the possible combinations. If none works, you know you're done. Doing it this way doesn't leave you wondering if you've missed anything.

In the next example, a sum is used.

Factor the quadratic: $2x^2 - 9x + 4$:

1. **Determine all the ways you can multiply two numbers to get 2.**

   You only have one choice: 2×1.

2. **Determine all the ways you can multiply two numbers to get 4.**

   $4 = 4 \times 1 = 2 \times 2$.

3. **The 4 is positive, so you want the sum of the outer and inner products.**

   To get a sum of 9, use the 2×1 and the 4×1 factors, multiplying $(2)(4)$ to get 8, and multiplying the two ones together to get 1. The sum of the 8 and the 1 is 9.

4. **Arrange your choices as binomials so the results are those you want.**

   $(2x\ 1)(1x\ 4)$

5. **Placing the signs, both binomials have to have subtraction so that the sum is −9 and the product is +4.**

   $(2x - 1)(1x - 4) = 2x^2 - 8x - 1x + 4 = 2x^2 - 9x + 4$

In the next example, all the terms are positive. The sum of the outer and inner products will be used. And you have several choices for the multipliers.

Factor: $10x^2 + 31x + 15$:

1. **Determine all the ways you can multiply two numbers to get 10.**

   The 10 can be written as 10×1 or 5×2.

2. **Determine all the ways you can multiply two numbers to get 15.**

   The 15 can be written as 15×1 or 5×3.

3. **The last term is +15, so you want the sum of the products to be 31.**

   Using the 5×2 and the 5×3, multiply (2)(3) to get 6, and multiply (5)(5) to get 25. The sum of 6 and 25 is 31.

4. **Arrange your choices in the binomials so the factors line up the way you want to give you the products.**

   $(2x\ 5)(5x\ 3)$

5. **Placing the signs is easy because everything is positive.**

   $(2x + 5)(5x + 3) = 10x^2 + 6x + 25x + 15 = 10x^2 + 31x + 15$

Coming to the end of the FOIL roll

The example in this section looks, at first, like a great candidate for factoring by this method. You'll see, though, that not everything can factor. Also, I get to make the point that using this method assures you that you've 'left no stone unturned' and can be confident when claiming that a trinomial is *prime* (can't be factored).

Factor: $18x^2 - 27x - 4$:

1. **Determine all the ways you can multiply two numbers to get 18.**

   The 18 can be written as 18×1 or 9×2 or 6×3.

2. **Determine all the ways you can multiply two numbers to get 4.**

   The 4 can be written as 4×1 or 2×2.

3. **Look at the sign of –4 and you see that you want a difference. And the difference of the products is to be 27.**

   You can't seem to find any combination that gives you a difference of 27. Run through all of them to be sure that you haven't missed anything.

Using the 18×1, cross it with:

- ✔ 4×1, which gives you a difference of either 14, using the (1)(4) and (18)(1), or 71, using the (1)(1) and the (18)(4).
- ✔ 2×2, which gives you a difference of 34, using (1)(2) and (18)(2); there's only one choice because both of the second factors are 2.

    Using the 9×2, cross it with:

- ✔ 4×1, which gives you a difference of either 34, using (2)(1) and (9)(4), or 1, using (2)(4) and (9)(1).
- ✔ 2×2, which gives you a difference of 14, only.

    Using the 6×3, cross it with:

- ✔ 4×1, which gives you a difference of either 21, using (3)(1) and (6)(4), or 6, using (3)(4) and (6)(1).
- ✔ 2×2, which gives you a difference of 6, only.

Because you've exhausted all the possibilities and you haven't been able to create a difference of 27, you can assume that this quadratic can't be factored. It's prime.

Making Factoring Choices

Sometimes you have to factor a problem more than once. This section shows you how you can use two completely different factoring techniques on the same problem. Note that the process of using different factoring techniques is different from reusing the same methods, such as taking out common factors several times.

Combining unFOIL and the greatest common factor

A quadratic, such as $40x^2 - 40x - 240$, can be factored using two different techniques, which can be done in two different orders. One of the choices makes the problem easier. It's the order in which the factoring is done that makes one way easier and the other way harder. You just have to hope that you recognise the easier way before you get started.

I show you the harder method first, so you'll see why it's important to make a good choice. In this example, the big numbers are left in and the unFOILing is done first.

Factor $40x^2 - 40x - 240$ by using FOIL first:

1. **Determine all the ways you can multiply two numbers to get 40.**

   The 40 can be written as 40×1, 20×2, 10×4, or 8×5.

2. **Determine all the ways you can multiply two numbers to get 240.**

   The 240 can be written as 240×1, 120×2, 80×3, 60×4, 48×5, 40×6, 30×8, 24×10, 20×12 or 16×15.

3. **The sign of the last term, -240, tells you that you want a difference. The difference of the products should be 40.**

   Using the 10×4 and the 20×12, multiply $(4)(20)$ to get 80, and multiply $(12)(10)$ to get 120. The difference between 80 and 120 is 40.

   Now, be honest with me. Did you find the listing of the multipliers and finding the right combination of cross-products to be a huge challenge? I certainly did!

4. **Arrange your choices as binomials and place the signs appropriately.**

   $$(4x - 12)(10x + 20) = 40x^2 - 40x - 240$$

But just look at the coefficients in those binomials. Each pair of coefficients can be factored itself. Each of the terms in the first binomial can be factored by 4, and each of the terms in the second binomial can be factored by 10.

$$(4x - 12)(10x + 20) = 4(x - 3)10(x + 2) = 40(x - 3)(x + 2)$$

Working this out took two types of factorisation — unFOILing and taking out a GCF.

For the next example, try the easier way. Factor out the GCF first.

Factor $40x^2 - 40x - 240$ by using the GCF first.

Each term's coefficient is evenly divisible by 40. Doing the factorisation:

$$40x^2 - 40x - 240 = 40(x^2 - x - 6)$$

Now, looking at the trinomial in the parentheses:

1. **Use unFOIL to factor the trinomial $x^2 - x - 6$.**

   The 1 that's the coefficient of the x^2 term can be written only as 1×1. The 6 can be written as 6×1 or 3×2. Notice how the list of choices is much shorter and more manageable than if you try to unFOIL before factoring out the GCF.

2. **Looking at the sign of the last term, –6, choose your products to create a difference of 1.**

   Using the 1×1 and the 3×2, it's easy to set up the factors:

   $$(1x \; 3)(1x \; 2)$$

   The middle term, x, is negative, so you want the $3x$, the product of the outer terms, to be negative. Finish the factoring. Then put the 40 that you factored out in the first place back into the answer.

   $$40x^2 - 40x - 240 = 40(x - 3)(x + 2)$$

You can get to the correct answer no matter what you choose to do in what order. As a general rule, though, factoring out a GCF first is usually best.

Grouping and unFOILing in the same package

With many types of factoring methods available to you in algebra, it's not surprising that you find so many different combinations of factorisations within a single problem. In the following section, you see factorisations of sums and differences of squares. Right now, I just stick to the methods covered so far in this chapter and in Chapter 8 to show you another interesting combination.

In the next example, you see six terms — a type of expression that seems to suggest using grouping to factor it. The big surprise comes after the grouping is finished.

Factor: $3x^2y - 24xy - 27y - 5x^2z + 40xz + 45z$.

The first three terms have coefficients divisible by 3, and each has a factor of y. So factor $3y$ out of each of those terms. The last three terms have coefficients that are divisible by 5, and each has a factor of z, so factor each term by $5z$.

$$= 3y(x^2 - 8x - 9) + 5z(-x^2 + 8x + 9)$$

You see that the two trinomials are not the same. In order for grouping to work, you have to create a fewer number of terms — each with some factor in common. The problem here is that I didn't factor out –5. Yes, I should've noticed, but I wanted to show you that *repairs* are easy. Changing the +5z to –5z, I factor –1 out of each term in the second trinomial and basically just change each sign to its opposite:

$$= 3y(x^2 - 8x - 9) - 5z(x^2 - 8x - 9)$$

Now I can factor the trinomial out of the two terms:

$$= (x^2 - 8x - 9)(3y - 5z)$$

The trinomial can be factored using unFOIL. You want the difference of the cross-products to be 8, so you use the factors 1×1 and 9×1:

$$= (x - 9)(x + 1)(3y - 5z)$$

Factoring the Difference of Two Perfect Squares

If two terms in a binomial are perfect squares and they're separated by subtraction, the binomial can be factored. A perfect square is not an old-school reference to a dancing partner with two left feet who refused to dance an entire evening. A perfect square is the result of multiplying a number or variable by itself. Twenty-five is a perfect square because it's equal to 5 times 5. To factor one of these binomials, just find the square roots of the two terms that are perfect squares and write the factorisation as the sum and difference of the square roots. For example, the difference of $x^2 - 49 = (x + 7)(x - 7)$. This rule only works if you're subtracting the squares. You can factor $x^2 - 9$ and $9 - x^2$, but you can't factor $x^2 + 9$ because that's a sum of squares, not a difference.

If subtraction separates two squared terms, the product of the sum and difference of the two square roots factors the binomial: $a^2 - b^2 = (a + b)(a - b)$.

For example, factor $9x^2 - 16$.

The square roots of $9x^2$ and 16 are $3x$ and 4, respectively. The sum of the roots is $3x + 4$ and the difference between the roots is $3x - 4$. So, $9x^2 - 16 = (3x + 4)(3x - 4)$.

Here are some more examples. Factor $25z^2 - 81y^2$.

The square roots of $25z^2$ and $81y^2$ are $5z$ and $9y$, respectively. So, $25z^2 - 81y^2 = (5z + 9y)(5z - 9y)$.

Factor $x^4 - y^6$.

The square roots of x^4 and y^6 are x^2 and y^3, respectively. So the factorisation of $x^4 - y^6 = (x^2 + y^3)(x^2 - y^3)$.

Factor $x^2 - 3$.

In this case, the second number is not a perfect square. But sometimes it's preferable to have the expression factored anyway. The square root of x^2 is x, and you can write the square root of 3 as $\sqrt{3}$ (For more on square roots and radicals, refer to Chapter 6.) Now the factorisation can be written: $(x - \sqrt{3})(x + \sqrt{3})$. Not pretty, but it's factored.

You may have noticed that I'm always writing the $(a + b)$ factor first and the $(a - b)$ factor second. It really doesn't matter in which order you write them; multiplication is commutative, so you can switch the factors, if you want. Just don't switch the terms in the $(a - b)$ factor.

Ending with binomials

In this section, I show you how you can start with four terms and apply a form of grouping (refer to the earlier section 'Grouping and unFOILing in the same package' if you need a refresher on that method). The result is the difference of two squares.

For example, factor $x^2 + 8x + 16 - y^2$.

When using grouping, you usually divide the four or six terms into equal-size groups. Sometimes four terms can be separated into unequal groupings with three terms in one group and one term in the other. The way to spot these special types of factoring situations is to look for squares. Of course, you usually don't even look for unequal groupings unless other grouping methods have failed you.

This expression has four terms, but no good equal pairing of terms will give you a set of useful common factors. Another option is to group unevenly. Group the first three terms together because they form a trinomial that can be factored. That leaves the last term by itself:

$$x^2 + 8x + 16 - y^2 = (x^2 + 8x + 16) - y^2$$

Now you can factor the trinomial in the parentheses using unFOIL:

$$(x + 4)^2 - y^2$$

Notice that you now have two terms and each is a perfect square.

Using the rule from the preceding section, $a^2 - b^2 = (a + b)(a - b)$, finish this example:

$$(x + 4)^2 - y^2 = [(x + 4) + y][(x + 4) - y]$$

Dropping the parentheses inside the brackets provides no big advantage, so leave the answer the way it is.

Knowing when to quit

One of my favourite scenes from the 1965 movie *The Agony and the Ecstasy*, which chronicles Michelangelo's painting of the Sistine Chapel, comes when the pope enters the Sistine Chapel, looks up at the scaffolding, dripping paint and Michelangelo perched up near the ceiling, and yells, 'When will it be done?' Michelangelo's reply: 'When I'm finished!'

The pope's lament can be applied to factoring problems: 'When is it done?'

Factoring is done when no more parts can be factored. If you refer to the listing of ways to factor two, three, four, or more terms, you can check off the options, discard those that don't fit, and stop when none works. After doing one type of factoring, you should then look at the values in parentheses to see if any of them can be factored.

For example, factor $x^4 - 104x^2 + 400$.

No GCF is possible, so the only other option with three terms is to unFOIL:

$$x^4 - 104x^2 + 400 = (x^2 - 4)(x^2 - 100)$$

You now have two factors, but each of them is the difference of perfect squares:

$$(x^2 - 4)(x^2 - 100) = (x + 2)(x - 2)(x + 10)(x - 10)$$

You're finished!

Here's another example. Factor $3x^5 - 18x^3 - 81x$.

The GCF of the terms is $3x$.

$$3x^5 - 18x^3 - 81x = 3x(x^4 - 6x^2 - 27)$$

The trinomial can be unFOILed:

$$3x(x^4 - 6x^2 - 27) = 3x(x^2 - 9)(x^2 + 3)$$

The first binomial is the difference of squares:

$$3x(x^2 - 9)(x^2 + 3) = 3x(x - 3)(x + 3)(x^2 + 3)$$

You're finished!

If you're really lucky your school may introduce you to CAS calculators in Years 9 and 10. All CAS calculators (doesn't matter which type you have) have a really cool function that allows you to expand or factorise equations and expressions. While this process is super easy, you still need to know the manual process, especially if you are heading towards Mathematical Methods or Specialist Maths and some of the wonderful advanced algebra and calculus that waits for you in those courses. Of course, become familiar with your CAS calculator by all means, and check your work with your CAS so that when you get to Years 11 and 12 using this technology is second nature.

Part III
Solving Algebraic Equations

Top Three Algebraic Equation Types

So you can keep them in mind as you go through this part, here's a list of the most common algebraic equation types and their most general format:

- **Linear equation:** $ax + b = c$
- **Quadratic equation:** $ax^2 + bx + c = 0$
- **Cubic equation:** $ax^3 + bx^2 + cx + d = 0$

In this part . . .

- Use investigation, logic and insight to solve problems — and perhaps entertain yourself at the same time.

- Solve problems featuring two terms, three terms, even fractions — and then try your hand at quadratic equations.

Chapter 10

Establishing the Ground Rules and Solving Linear Equations

. . . .

In This Chapter

▶ Recognising where to start when solving equations

▶ Keeping your balance while performing equation manipulations

▶ Reconciling your work and understanding its practical applications

▶ Finding the most direct path to your solution

▶ Solving equations with two or three terms

▶ Considering alternate methods of dealing with fractions

▶ Adapting formulas to your needs and preferences

. . . .

*I*n this chapter, you find many different considerations involving solving equations in algebra. In earlier chapters, I cover the mechanics of working with algebraic expressions correctly. Now I put those rules to work by introducing the equal sign (=). Just as a verb makes a phrase into a sentence, an equal sign makes an expression into an equation. In this chapter, instead of dealing with expressions, such as $3x + 2$, I show you how to prepare for solving equations, such as $3x + 2 = 11$.

I also cover linear equations, which consist of some terms that have variables and others that are constants. A standard form of a linear equation is $ax + b = c$. What distinguishes linear equations from the rest of the pack is the fact that the variables are always raised to the first power (Chapter 11 covers solving quadratic equations).

In this chapter, I take you through many different types of opportunities for dealing with linear equations. Most of the principles you use with these first-degree equations are applicable to the higher-order equations, so you don't have to start from scratch later on.

When you use algebra in the real world, more often than not you turn to a formula to help you work through a problem. Fortunately, when it comes to algebraic formulas, you don't have to reinvent the wheel: You can make use of standard, tried-and-true formulas to solve some common, everyday problems. I show you how to change the format or adapt many of your favourite formulas to make them more usable for your particular situation.

Creating the Correct Setup for Solving Equations

Different types of equations take different types of handling in order to solve for the correct solution, all the solutions, and not too many solutions. The setup of the equation depends on which type of equation you're dealing with at that time.

Here's a list of the most common algebraic equation types and their most general format:

- **Linear equation:** $ax + b = c$
- **Quadratic equation:** $ax^2 + bx + c = 0$
- **Cubic equation:** $ax^3 + bx^2 + cx + d = 0$

The convention is to use the letters toward the beginning of the alphabet as numbers or constants and the letters toward the end of the alphabet for variables.

In general, equations are usually set equal to 0 or some constant. I cover the differences and similarities in detail in the next few chapters.

Keeping Equations Balanced

When presented with an algebraic equation, your usual task is to solve the equation. *Solving* an equation means to find the value or values that replace the unknown(s) to make the equation a true statement. You may be able just to guess the answer, but you can't rely on that method for all equations. Some answers are just too hard to find by guessing.

In this section, I tell you about the tried-and-true, approved methods of changing the original equation so that it's in a format that shows you the solution.

Balancing with binary operations

One of the most efficient and easiest methods of changing the format of an equation is to perform arithmetic operations on each side of the equation. Picture a seesaw or balance scale. When the seesaw is in balance, the two ends are at the same level and the board is parallel to the ground. With a balance scale, you weigh items by placing the object in one tray and then adding known weights to the other tray until the two are in balance. Algebraic equations start out in balance, and your task is to keep them that way.

Adding to each side or subtracting from each side

When changing the format of an equation, if you add some amount (or subtract some amount) from one side of the equation, you must do the same thing to the other side.

Here are some examples of adding or subtracting from each side of the equation:

- If $2x - 10 = 46$, then $2x - 10 + 10 = 46 + 10$, or $2x = 56$
- If $x^2 + 5x = 10$, then $x^2 + 5x - 10 = 10 - 10$, or $x^2 + 5x - 10 = 0$
- If $\sqrt{5x - 2} - 10 = x$, then $\sqrt{5x - 2} - 10 + 10 = x + 10$, or $\sqrt{5x - 2} = x + 10$

Multiplying each side by the same number

You can multiply each side of an equation by any number and not change the equality of the statement. You can even multiply each side by 0 and not change the equality (because you'd have $0 = 0$), but you wouldn't have much to work with if you did that. And you can change a false statement into a true statement by multiplying each side by 0. That's not playing fair.

Say you have the equation $\frac{3x}{10} + 2 = x$. Multiply both sides of the equation by 10 to solve it. **Remember:** You have to multiply each of the three terms by 10.

$$10\left(\frac{3x}{10}+2\right)=10(x)$$

$$\cancel{10}\left(\frac{3x}{\cancel{10}}\right)+10(2)=10x$$

$$3x+20=10x$$

Dividing each side of the equation by the same number

An algebraic equation remains balanced — it has the same solution as the original — when you divide each side by the same number. The one exception to the rule is dividing by 0; you just can't do that.

Usually, you won't be tempted to divide by 0, but you might do it accidentally. For example, you might decide to divide each side of an equation by the binomial $x - 2$. Seems harmless enough, but, if the value of x is 2, you've divided by 0 and created a non-answer. Also, dividing by 0 can cause you to lose a solution.

Say you're faced with the following equation: $10\sqrt{x+3}-8=40$. Divide both sides of the equation by 10:

$$\frac{\cancel{10}\sqrt{x+3}}{\cancel{10}}-\frac{\cancel{8}^{4}}{\cancel{10}_{5}}=\frac{\cancel{40}^{4}}{\cancel{10}}$$

$$\sqrt{x+3}-\frac{4}{5}=4$$

Every single term gets divided by the selected number.

As another example, solve the equation $x^2 = 4x$.

Divide both sides of the equation by x. Hold on! This is the situation I warned about. In the original equation, you see two different answers or solutions for the equation — that is, x can be 4, giving you $16 = 16$, or x can be 0, making the equation $0 = 0$. Either number makes the equation true. But, if you divide each side by x, you lose the solution that $x = 0$:

$$\frac{x^{\cancel{2}^{1}}}{\cancel{x}}=\frac{4\cancel{x}}{\cancel{x}}$$

$$x = 4$$

Squaring both sides and suffering the consequences

Some equations require squaring terms in order to solve them. The obvious candidates for squaring are equations containing radicals. Squaring both sides of an equation is a useful technique, but it comes with some concerns. Deal with the concerns, and you're fine. Let me show you what happens, and when to watch out for the tricky part.

Consider, for example, the equation $\sqrt{8} = \sqrt{x}$. If I square both sides, the radicals disappear: $(\sqrt{8})^2 = (\sqrt{x})^2$, giving me $8 = x$. Works for me! So what's the big problem?

Look at this next example: I start with $5 = -5$. The statement is false, but, if I square both sides, I make the statement true: $(5)^2 = (-5)^2$ gives me $25 = 25$. You say that you'd never do such a thing — take an untrue statement and make it appear to be true. But you just might do that inadvertently if the original equation contains a variable.

For example, if I start with the equation $x = \sqrt{100}$ and square both sides, I get the equation $x^2 = 100$. Two numbers make the second equation true — both 10 and -10. But the -10 isn't a solution of the original equation; $-10 \neq \sqrt{100}$, because the square root of 100 is $+10$, only.

Using a balance scale to find the counterfeit coin

You're given nine identical-looking gold coins and told that one of the coins is counterfeit. The counterfeit coin weighs slightly more than the real coins, but not enough for you to tell just by holding the coins in your hands. You have a balance scale and you're allowed to use it just twice. How can you determine which is the counterfeit coin with just two weighings? (Think about it, before reading on for the answer.)

You divide the coins into three piles of three coins each. Put one pile in one tray of the balance scale and a second pile in the other tray. If one side is lower, the counterfeit coin must be in that pile. If the two sides of the balance scale are the same, the counterfeit coin must be in the pile not on the scale.

After you've determined the pile of three that contains the counterfeit coin, put two of the coins on the balance scale (one on each tray). (This is your second weighing.) If one side is lower, that's your counterfeit. If they're the same, the coin not on the scale is counterfeit.

Taking a root of both sides

The opposite of squaring both sides of an equation is taking a root of both sides. If your variable is raised to the second power, you take a square root. If the variable is raised to the third power, you take a third root, and so on.

In the preceding section, I tell you that $\sqrt{100}$ is equal to +10 only. When you start out with an equation containing a radical like this one, you only consider the positive answer. But — and here's the catch — when you start out with a second power and *take the square root*, you can consider both positive and negative answers.

Here are some examples of taking the square root of both sides of an equation:

- $x^2 = 49$: Taking the square root, you put $\pm$ in front of the radicals to show that both a positive and negative number can give you that value: $\pm\sqrt{x^2} = \pm\sqrt{49}$, giving you $\pm x = \pm 7$. Putting the $\pm$ on both sides is really overkill. If both sides are positive, you get $x = 7$. If both sides are negative, you have $-x = -7$, which is the same as $x = 7$ after multiplying both sides by -1. If the left side is positive and the right side negative, you get $x = -7$. If it's the opposite, you get $-x = 7$, which is the same as $x = -7$, multiplying both sides by -1. So, by convention, you put the $\pm$ on just one side of the equation.

  After all that, did you forget what I was doing? If so, let me just tell you that the answer is $x = \pm 7$, which means that $x = 7$ or $x = -7$.

- $64 = (x + 1)^2$: Taking the square root of each side, you get $\pm\sqrt{64} = \sqrt{(x+1)^2}$ or that $\pm 8 = x + 1$. (I put the $\pm$ on just one side.) The two numbers that work in this equation are 7 and –9.

Undoing an operation with its opposite

One helpful method used when working with equations is adding and subtracting the same thing to one side or the other — or multiplying and dividing one side by the same thing. Look at the following equations:

- If $x = 8$, then $x + 2 - 2 = 8$.

- If $y = 11$, then $\frac{6y}{6} = 11$.

You're probably looking at these equations and saying, 'Well, that did a lot of good!' And I don't blame you. I show you this now because I'm stressing keeping equations balanced.

Solving with Reciprocals

Multiplication and division are opposite operations. Multiplication is undone by division and vice versa, as I explain in the previous sections. Another option, though, may work better in certain circumstances — using the reciprocal, or multiplicative inverse, of the number that you're trying to 'get rid of'. Choose this alternative if a fraction is multiplying the variable, such as in $\frac{3x}{19}=12$.

Two numbers are reciprocals if multiplying them together yields a product of 1.

Look at the following examples of reciprocals:

- ✔ 5 and $\frac{1}{5}$ are reciprocals of each other: $5\left(\frac{1}{5}\right)=1$.
- ✔ $-\frac{3}{7}$ and $-\frac{7}{3}$ are reciprocals: $\left(-\frac{3}{7}\right)\left(-\frac{7}{3}\right)=1$.
- ✔ The reciprocal of a is $\frac{1}{a}$ (as long as a isn't 0).
- ✔ The reciprocal of $\frac{1}{b}$ is b (as long as b isn't 0).

Solving equations in the fewest possible steps is usually preferable. Multiplying by the reciprocal replaces the two steps of multiplying each side and dividing each side. That's why you can choose to multiply both sides of an equation by $\frac{5}{4}$, the reciprocal of $\frac{4}{5}$, to solve for a in the expression $\frac{4a}{5}$, which can be thought of as $\left(\frac{4}{5}\right)a$.

In the following examples, both sides of the equation are multiplied by the reciprocal of the fraction multiplying the variable.

In this example, the variable is being multiplied by $\frac{4}{5}$:

$$\frac{4a}{5}=12$$

Multiply each side by the reciprocal, $\frac{5}{4}$:

$$\frac{5}{4}\left(\frac{4a}{5}\right)=\left(\frac{5}{4}\right)\cdot 12$$

Reduce and simplify:

$$\frac{\cancel{5}}{\cancel{4}}\left(\frac{\cancel{4}a}{\cancel{5}}\right)=\left(\frac{5}{4}\right)\cdot \cancel{12}^{\;3}$$

$$a = 15$$

As another example, solve $\frac{x}{2} = 19$ for x.

$\frac{x}{2}$ is another way of saying $\left(\frac{1}{2}\right)x$. So you can solve by multiplying by the reciprocal of $\frac{1}{2}$, which is 2:

$$\frac{\cancel{2}}{1}\left(\frac{1x}{\cancel{2}}\right) = 19\left(\frac{2}{1}\right)$$

$$x = 38$$

Solve $-f = 11$ for f. This is an easy equation to solve, but you may be surprised at how many people get the wrong answer — all because of a little dash in front of a letter. Think of the f as being multiplied by -1. Putting in the -1 gives you a multiplier that you can work with to solve the equation. What's the reciprocal of -1? It's -1!

$$-f = 11$$

$$(-1)(-1f) = 11(-1)$$

$$f = -11$$

Another example involves using the reciprocal of a fraction, even though it doesn't look like it at first.

Solve for x in $0.7x = 42$.

One way to solve this equation is to divide each side by 0.7.

A decimal point can get lost easily when it's in the front of a term. You may miss it or think it's a fly speck. Putting a 0 in front of a decimal point draws attention to the decimal and doesn't change the value of the number. Look at the difference between .8 and 0.8 in this sentence.

To solve the equation using a reciprocal, first convert 0.7 to $\frac{7}{10}$, thereby replacing the decimal with the fraction. The reciprocal of $\frac{7}{10}$ is $\frac{10}{7}$.

$$\frac{7}{10}x = 42$$

$$\frac{\cancel{10}}{\cancel{7}}\left(\frac{\cancel{7}}{\cancel{10}}\right)x = 42\left(\frac{10}{7}\right)$$

$$x = \frac{\cancel{420}}{\cancel{7}}$$

$$x = 60$$

Making a List and Checking It Twice

Algebraic problems can be naughty or nice — or something in between. Whatever the case, you want to have confidence in your work and get good results. When you're performing operations in algebra, and solving equations, you want to check your work. You can use the careful, methodical mechanical check of your processes and numbers, or the reality check.

I do the reality check first because it's usually a visual or quick take on whether the answer fits the situation. Mechanical checks can be tricky. I don't know about you, but I often make the same mistake all over again when I check my work too quickly after doing it.

When possible, I try to leave at least half an hour between doing the original work and checking it. That's not always possible — especially on a test — but it usually works better for me.

Doing a reality check

Reality TV and reality checks in algebra: What do they have in common? Actually, they have very little in common. Reality checks in algebra are much more believable than reality TV is.

I usually do a reality check by asking: Does this answer make any sense? Would there really be 1,000 doughnuts in the paper bag? Does the sum really come out to be a negative number? If the answer doesn't make sense, chances are you've made a computational error. Go back and try again.

For example, say the number of soccer players participating at a summer soccer camp is 330, with 11 players from each club. You're preparing club participation certificates to give to each club captain, so you need to know: How many clubs are represented?

To show you that a reality check can save you from making a big error, pretend that you didn't really think this through and decided to solve the problem with the following equation:

$$\frac{c}{11} = 330$$

The letter *c* represents the number of soccer clubs. You divide *c* by the number of players in each club and set it equal to the total number of players.

You used the variable and the two numbers in the problem. Does it matter what you use where? Will the equation give you a reasonable answer?

Multiply each side by 11 to solve for *c*:

$$11 \cdot \frac{c}{11} = 330 \cdot 11$$

$$\cancel{11} \cdot \frac{c}{\cancel{11}} = 330 \cdot 11$$

$$c = 3{,}630 \text{ clubs}$$

Humph. This can't be right. The answer doesn't make any sense — only 330 players are involved. The answer may satisfy your equation, but if it doesn't make sense, the equation could be wrong.

A quick look at the equation shows that it should have read:

11 players per club × the number of clubs = the total number of players

$$11c = 330$$

Now, solve this:

$$11c = 330$$

Divide each side by 11:

$$\frac{11}{11}c = \frac{330}{11}$$

$$\frac{\cancel{11}}{\cancel{11}}c = \frac{330}{11}$$

$$c = 30 \text{ clubs}$$

That makes much more sense.

You can solve an equation correctly, but that doesn't mean you chose the right equation to solve in the first place. Make sure that your answer makes sense.

Thinking like a car mechanic when checking your work

The more complete check of algebraic processes is checking the computations and algebraic operations. When a car mechanic has a spiffy-doodle computer to run a diagnostic check on your car, she finds all the problems very efficiently. If your car is older, though, or she doesn't have that kind of electronic setup, she needs to perform a step-by-step, point-by-point check of all the essential parts. This is more like an algebraic check.

I want to see how good you are at checking work. Here's a problem that a student did, and the answer is wrong. It's more helpful to someone who's made an error when you can point out where the error is in his computations. Can you find it? (The answer is –2.)

$$\frac{-6\left[3^2 + 4 - 5(2)\right]}{\sqrt{16} + 5} = \frac{-6\left(3^2\right) - 6(4) - 6\left[5(2)\right]}{4 + 5} = \frac{-6(9) - 6(4) - 6(10)}{9}$$

$$= \frac{-54 - 24 - 60}{9} = -\frac{138}{9} = -\frac{46}{9}$$

You probably spotted the error right away because one of the most common mistakes in distributing is in not distributing the negative signs correctly. Yes, you're right, the third term in the top-right fraction should be –6[–5(2)]. (I show you the distribution of signed numbers in Chapter 3, if you need a refresher.)

Finding a Purpose

One of the questions asked most frequently by students who are either tired of the algebra homework or frustrated by the challenges is: 'When will I ever use this?' My answer usually depends on the situation, but sometimes it's hard to come up with a convincing response.

When you're studying algebra so that you can be successful in calculus, there's really no question as to the why and when you'll use it. And, believe it or not, some people study algebra for the pure joy of doing so. (Are you one of those special people?) But, in this section, I really should give an example of how you could apply algebra — operations and variables and symbols — to organise and solve a problem.

Say a famous group sold 130,000 copies of its latest CD. This particular CD cost $26. If the group's share was $845,000, what was its percentage cut of the gross sales amount?

Letting p represent the percentage of the total revenue and using the 130,000 and $26 per DVD, you can set up the following equation:

Income = (number DVDs)(price per DVD)(percentage)

$845,000 = (130,000)(26)(p)$

Solving the equation for p, you divide each side by the product of 130,000 and 26. (I show you how to solve this type equation in the following sections in this chapter.)

$$\frac{845,000}{(130,000)(26)} = p$$

The value of p comes out to be 0.25 or 25 per cent.

Solving Linear Equations: Playing by the Rules

When you're solving equations with just two terms or three terms or even more than three terms, the big question is, 'What do I do first?'

Actually, as long as the equation stays balanced, you can perform any operations in any order. But you also don't want to waste your time performing operations that don't get you anywhere or even make matters worse.

The following list tells you how to solve your equations in the best order. The basic process behind solving equations is to use the *reverse* of the order of operations.

The order of operations (refer to Chapter 5) is powers or roots first, then multiplication and division, and addition and subtraction last. Grouping symbols override the order. You perform the operations inside the grouping symbols to get rid of them first.

So, reversing the order of operations:

1. **Do all the addition and subtraction.**

   Combine all terms that can be combined both on the same side of the equation and on opposite sides using addition and subtraction.

2. **Do all multiplication and division.**

   This step is usually the one that isolates or solves for the value of the variable or some power of the variable.

3. **Multiply exponents and find the roots.**

   Powers and roots aren't found in these linear equations — they come in quadratic and higher-powered equations. But these would come next in the reverse order of operations.

When solving linear equations, the goal is to isolate the variable you're trying to find the value of. Isolating it, or getting it all alone on one side, can take one step or many steps. And it has to be done according to the rules — you can't just move things willy-nilly, helter-skelter, hocus pocus ... you get the idea.

Solving Equations with Two Terms

Linear equations contain variables raised to the first power. The easiest types of linear equations to solve are those consisting of just two terms. The following equations are all examples of linear equations in two terms:

$$14x = 84 \qquad -64 = 8y \qquad \qquad \frac{9z}{5} = 18 \qquad \qquad \frac{7w}{6} = \frac{35}{9}$$

Linear equations that contain just two terms are solved with multiplication, division, reciprocals or some combinations of the operations.

Devising a method using division

One of the most basic methods for solving equations is to divide each side of the equation by the same number. Many formulas and equations include a *coefficient* (multiplier) with the variable. To get rid of the coefficient and solve the equation, you divide. The following example takes you step by step through solving with division.

Solve for x in $20x = 170$.

1. **Determine the coefficient of the variable and divide both sides by it.**

   Because the equation involves multiplying by 20, undo the multiplication in the equation by doing the opposite, which is division. Divide each side by 20:

   $$\frac{20x}{20} = \frac{170}{20}$$

2. **Reduce both sides of the equal sign.**

   $$\frac{\cancel{20}x}{\cancel{20}} = \frac{170}{20}$$

   $$x = 8.5$$

Do unto one side of the equation what the other side has had done unto it.

Next, I show you two examples with practical applications embedded in them.

Your teacher needs to buy 300 doughnuts for a big parent meeting. How many dozen doughnuts is that?

Let d represent the number of dozen doughnuts you need. A dozen has 12 doughnuts, so $12d = 300$. Twelve times the number of doughnuts you need has to equal 300.

1. **Determine the coefficient of the variable and divide both sides by it.**

   Divide each side by 12.

   $$\frac{12d}{12} = \frac{300}{12}$$

2. **Reduce both sides of the equal sign.**

   $$d = 25 \text{ dozen doughnuts}$$

Here's another example. The display board at the bank says your parents could earn $4\frac{7}{8}$ per cent interest on their investment. (They're saving up for a family holiday.) They'd like to have earnings of at least $500 over the next year. How much do they have to deposit with the bank at that rate?

Write this financial puzzle as an equation, letting x represent the amount of money your parents need to invest. You get the amount of interest by multiplying the *principal* (amount invested) times the interest rate (written as a decimal). The equation you need is: $500 = x(0.04875)$.

The decimal value 0.04875 multiplies the variable x, so divide each side of the equation by that decimal number:

$$\frac{500}{0.04875} = \frac{x\left(\cancel{0.04875}\right)}{\cancel{0.04875}}$$
$$10{,}256.41026 = x$$

You'd have to invest about \$10,260 to earn the \$500.

This example doesn't include any provision for *compound interest* (where the interest is figured more than once a year and the earned amount is added to the principal). So, technically, you'd end up with more than \$500 at the end of the year.

Making the most of multiplication

The opposite operation of multiplication is division. I use division in the preceding section to solve equations where a number multiplies the variable. The reverse occurs in this section: I use multiplication where a number already divides the variable. The first example walks you through the steps needed.

Solve for y in $\frac{y}{11} = -2$.

1. **Determine the value that divides the variable and multiply both sides by it.**

   In this case, 11 is dividing the y, so that's what you multiply by.

   $$11\left(\frac{y}{11}\right) = (-2)(11)$$

2. **Reduce both sides of the equal sign.**

   $$\cancel{11}\left(\frac{y}{\cancel{11}}\right) = -22$$
   $$y = -22$$

Next, look at an example that's applicable — a bit hairy (pardon the pun), but you read about situations like this all the time.

A wealthy woman's will dictated that her fortune be divided evenly among her nine cats. Each feline got $500,000, so what was her total fortune before it was split up? (Cats don't pay tax. Does that give you paws? Ouch.)

Let f represent the amount of her fortune. Then you can write the equation:

$$\frac{f}{9} = 500,000$$

In other words, the fortune divided by 9 gave a share of $500,000.

1. **Determine the value that divides the variable and multiply both sides by it.**

   In this equation, the fortune was divided. Solve the puzzle by multiplying each side by 9. The opposite of division is multiplication, so multiplication undoes what division did.

   $$9\left(\frac{f}{9}\right) = 500,000 \cdot 9$$

2. **Reduce on the left and multiply on the right.**

   $$9\left(\frac{f}{9}\right) = 4,500,000$$

   $$f = \$4,500,000$$

Her fortune was $4.5 million! Those are nine very happy kitties. You can bet their caretakers hope they have nice, long lives.

In the next example, the variable is both multiplied by 4 and divided by 5. You solve the problem using both multiplication and division.

Solve for a in $\frac{4a}{5} = 12$.

1. **Determine what is dividing the variable.**

   In this case, the 5 is dividing both the 4 and the variable a.

2. **Multiply the values on each side of the equal sign by 5.**

   $$5\left(\frac{4a}{5}\right) = 12(5)$$

3. Reduce and simplify.

$$5\left(\frac{4a}{5}\right) = 12(5)$$

$$4a = 60$$

4. Determine what is multiplying the variable.

The number 4 is the coefficient and multiplies the a.

5. Divide the values on each side of the equal sign.

$$\frac{4a}{4} = \frac{60}{4}$$

6. Reduce and simplify.

$$\frac{4a}{4} = \frac{60}{4}$$

$$a = 15$$

A simpler way of solving this last equation is to multiply by the reciprocal of the variable's coefficient. I show you that alternative next.

Reciprocating the invitation

The reciprocal of a number is its 'flip'. A more mathematical definition is that a number and its reciprocal have a product of 1.

What makes the reciprocal so important in algebra is that you can create the number 1 as a coefficient of a variable by multiplying by the reciprocal of the current coefficient. So, in a way, this process is just a special case of multiplying each side by the same number.

In the first example that follows, I solve a problem (the last example in the preceding section) using the reciprocal rather than doing the two operations of multiplication and division.

Solve for a in $\frac{4a}{5} = 12$.

The coefficient of the variable a is the fraction $\frac{4}{5}$. The reciprocal of $\frac{4}{5}$ is $\frac{5}{4}$. So, to solve for a, you multiply each side of the equation by $\frac{5}{4}$:

$$\frac{\cancel{5}}{\cancel{4}} \cdot \frac{\cancel{4}a}{\cancel{5}} = \frac{\overset{3}{\cancel{12}}}{1} \cdot \frac{5}{\cancel{4}}$$

$$a = 15$$

Extending the Number of Terms to Three

The standard form of a linear equation is $ax + b = c$. In the 'Solving Equations with Two Terms' section, earlier in this chapter, you have linear equations for which the value of b is 0, which gives you just $ax = c$. In this section, I introduce that extra constant value and show you how to deal with it. Also, in this section, you find equations that start out with more than one variable term, and you work toward combining and creating a new equation with just the one variable.

In general, you solve linear equations by simplifying and performing operations that give you a variable term on one side of the equal sign and a constant term on the other side. Then you can use multiplication or division to finish the problem.

Eliminating the extra constant term

When you have a linear equation involving three terms, and just one of the terms contains a variable factor, you add or subtract a constant to isolate that variable term — get it by itself on one side of the equation.

For example, solve for y in $3y - 11 = 19$.

To isolate the y term, you add 11 to each side of the equation. The number 11 is chosen, because it's the opposite of –11, and the sum of –11 and 11 is 0.

$$
\begin{array}{rl}
3y - 11 = & 19 \\
+11 & +11 \\
\hline
3y = & 30
\end{array}
$$

Now you have a linear equation in two terms, which is solved by dividing each side of the equation by 3:

$$\frac{\cancel{3}y}{\cancel{3}} = \frac{\overset{10}{\cancel{30}}}{\cancel{3}}$$

$$y = 10$$

In the next example, some of the characters seem to be out of order, but the properties of algebra come into play and allow you to use the same processes, no matter how the problem starts out.

Solve for z in $41 = 14 - 9z$.

First, get the variable term by itself on the right side by subtracting 14 from each side of the equation:

$$\begin{array}{r} 41 = 14 - 9z \\ \underline{-14 - 14} \\ 27 = \quad -9z \end{array}$$

Now divide each side of the equation by -9 to solve for z:

$$\frac{\overset{3}{\cancel{27}}}{-\cancel{9}} = \frac{\cancel{-9}z}{\cancel{-9}}$$

$$-3 = z$$

Another way of writing the solution is $z = -3$.

Some people prefer working with the variable term on the left — it's just what they're used to. No problem. At any point in your work, you can switch the sides. So, starting at the beginning, the equation $41 = 14 - 9z$ becomes $14 - 9z = 41$. You notice that I don't change the order of the terms — just the sides that they lie on.

The *symmetric property* of equations says that if $a = b$, then $b = a$.

Vanquishing the extra variable term

One aim of the linear-equation solver is to get the variable term on one side of the equation and the constant term on the other side. In the preceding section, I show you how to get rid of the pesky extra constant. But what if you have more than one variable term? Can that be dealt with as easily as the constant numbers? The answer is a resounding 'Yes.'

To reduce your linear equation to one variable term, you first perform any addition or subtraction necessary to get all the variable terms on one side of the equation. Then you combine those variable terms in the same manner that I show you in Chapter 7.

For example, solve for the value of x in $5x - 4 = 3x + 8$.

First, subtract $3x$ from each side of the equation. That step removes the x term from the right side. Subtracting $5x - 3x$, you get $2x$ because the two terms have the same variable:

$$\begin{aligned} 5x - 4 &= 3x + 8 \\ -3x \quad &\quad -3x \\ \hline 2x - 4 &= \quad 8 \end{aligned}$$

Now the problem looks like those from the previous section. I isolate the variable term by adding 4 to each side of the equation:

$$\begin{aligned} 2x - 4 &= \ 8 \\ +4 \quad &\ +4 \\ \hline 2x \quad &= 12 \end{aligned}$$

Now the problem is finished by dividing each side of the equation by 2:

$$\frac{\cancel{2}x}{\cancel{2}} = \frac{\overset{6}{\cancel{12}}}{\cancel{2}}$$

$$x = 6$$

In this example, solve for w in $3w - 5 + 4w = 16 - w + 3$.

You see two variable terms on the left side of the equation and two constant terms on the right side. Combine those terms first:

$$-5 + 7w = 19 - w$$

Add w to each side of the equation; then add 5 to each side:

$$\begin{array}{r} -5+7w = 19-w \\ +w \qquad +w \\ \hline -5+8w = 19 \\ +5 \qquad +5 \\ \hline 8w = 24 \end{array}$$

Now you can divide each side of the equation by 8 and get $w = 3$.

Simplifying to Keep It Simple

Linear equations don't always start out in the nice $ax + b = c$ form. Sometimes, because of the complexity of the application, a linear equation can contain multiple variable and constant terms and lots of grouping symbols, such as in this equation:

$$3[4x + 5(x + 2)] + 6 = 1 - 2[9 - 2(x - 4)]$$

The different types of grouping symbols are used for nested expressions (one inside the other), and the rules regarding order of operations (refer to Chapter 5) apply as you work toward figuring out what the variable x represents.

Distributing first

Distributing means that the number or variable next to the grouping symbol multiplies every value inside the grouping symbol. Equations containing grouping symbols offer opportunities for making wise decisions. In some cases you need to distribute, working from the inside out, and in other cases it's wise to multiply or divide first (see the following section). In general, you'll distribute first if you find more than two terms in the entire equation.

For example, solve for y in $8(3y - 5) = 9(y - 6) - 1$.

The equation has two terms involving grouping symbols. Distribute the 8 and 9, first:

$$24y - 40 = 9y - 54 - 1$$

Combine the two constant terms on the right. Then subtract $9y$ from each side of the equation:

$$\begin{aligned} 24y - 40 &= 9y - 55 \\ -9y &\quad -9y \\ \hline 15y - 40 &= \quad -55 \end{aligned}$$

Now add 40 to each side of the equation; then divide each side by 15:

$$\begin{aligned} 15y - 40 &= -55 \\ +40 &\ +40 \\ \hline 15y &= -15 \\ \frac{15y}{15} &= \frac{-15}{15} \\ y &= -1 \end{aligned}$$

Now let me show you the solution of the example I give at the beginning of this section.

Solve for x in $3[4x + 5(x + 2)] + 6 = 1 - 2[9 - 2(x - 4)]$.

The best way to sort through all these operations is to simplify from the inside out. You see parentheses within brackets. The binomials in the parentheses have multipliers. I'll step through this carefully to show you an organised plan of attack.

First, distribute the 5 over the binomial inside the left parentheses and the -2 over the binomial inside the right parentheses:

$$3[4x + 5x + 10] + 6 = 1 - 2[9 - 2x + 8]$$

Now combine terms within the brackets:

$$3[9x + 10] + 6 = 1 - 2[17 - 2x]$$

Distribute the 3 over the two terms in the left brackets and the -2 over the terms in the right brackets:

$$27x + 30 + 6 = 1 - 34 + 4x$$

The constant terms on each side can be combined:

$$27x + 36 = -33 + 4x$$

Now subtract $4x$ from each side and subtract 36 from each side:

$$\begin{aligned} 27x + 36 &= -33 + 4x \\ -4x \qquad\quad\ &\ \ -4x \\ \hline 23x + 36 &= -33 \\ -36 &\ -36 \\ \hline 23x \qquad &= -69 \end{aligned}$$

Now, dividing each side of the equation by 23, you get that $x = -3$.

Multiplying or dividing before distributing

In this section, I show you where it might be easier to divide through by a number rather than distribute first. My only caution is that you always divide (or multiply) each term by the same number.

For example, solve for z in $12z - 3(z + 7) = 6(z - 1)$.

In this equation, you see three terms: Two on the left and one on the right. Each term has a multiplier of a multiple of 3. So divide each term by 3:

$$\frac{^4\cancel{12}z}{\cancel{3}} - \frac{\cancel{3}(z+7)}{\cancel{3}} = \frac{^2\cancel{6}(z-1)}{\cancel{3}}$$
$$4z - (z+7) = 2(z-1)$$

Notice that the second term has the negative sign in front of the resulting binomial. Be very careful not to lose track of the negative multipliers.

Distribute the negative sign and the 2:

$$4z - z - 7 = 2z - 2$$

Combine the two variable terms on the left. Then subtract $2z$ from each side:

$$\begin{aligned} 3z - 7 &= 2z - 2 \\ -2z \qquad &\ -2z \\ \hline z - 7 &= \quad -2 \end{aligned}$$

Finally, add 7 to each side and you get:

$$z = -2 + 7 = 5$$

The next example mixes two different situations that are actually the same. The terms in the equation either have a fractional multiplier or are in a fraction themselves. The point of the example is to show when multiplying each term by the same number first is preferable to distributing first.

Solve for x in the following equation:

$$\frac{3(x-2)}{4} + \frac{1}{2}(5x+2) = \frac{14x+12}{8} + 7$$

At first glance, the equation looks a bit forbidding. But quick action — in the form of multiplying each term by 8 — takes care of all the fractions. You're left with rather large numbers, but that's still nicer than fractions with different denominators. I choose to multiply by 8 because that's the least common denominator of each term (even the last term). Each of the four terms is multiplied by 8:

$$\frac{{}^2\cancel{8}}{1}\left[\frac{3(x-2)}{\cancel{4}}\right] + \frac{{}^4\cancel{8}}{1}\left[\frac{1}{\cancel{2}}(5x+2)\right] = \frac{\cancel{8}}{1}\left[\frac{14x+12}{\cancel{8}}\right] + 8(7)$$

$$6(x-2) + 4(5x+2) = (14x+12) + 56$$

Do the multiplication and distribution in steps to avoid errors:

$$6x - 12 + 20x + 8 = 14x + 12 + 56$$

The two variable terms on the left and the two constant terms on the left can be combined. Likewise, combine the two constant terms on the right:

$$26x - 4 = 14x + 68$$

Now subtract $14x$ from each side and add 4 to each side:

$$\begin{aligned}
26x - 4 &= 14x + 68 \\
-14x \quad\quad &\quad -14x \\
\hline
12x - 4 &= \quad\quad 68 \\
+4 \quad\quad &\quad\quad +4 \\
\hline
12x \quad &= \quad\quad 72
\end{aligned}$$

Dividing each side of the equation by 12, you see that $x = 6$.

When eliminating fractions in an equation, you need to multiply through by the least common denominator of all the fractions in the terms. In the preceding example, the least common denominator was 8. In the equation $\frac{x}{6}+\frac{x}{12}+\frac{x}{15}=1$, the least common denominator is 60.

The common denominator for fractions with denominators of 6, 12 and 15 is 60. How did I get this? One way is to guess. Another way is to write the prime factorisations of the numbers and find what they're all common to. Another quick trick is given in the following steps.

1. **Find the least common denominator by taking the biggest of the denominators and checking all its multiples until you find one that all the denominators divide.**

   In the case of this problem, 15 is the biggest denominator:

   - $15 \times 1 = 15$: Neither 6 nor 12 divides that evenly.

   - $15 \times 2 = 30$: Only the 6 divides that evenly.

   - $15 \times 3 = 45$: Neither 6 nor 12 divides that evenly.

   - $15 \times 4 = 60$: A winner!

2. **Multiply each fraction by that common denominator.**

   When you multiply by 60 in the sample problem, all the denominators divide out or disappear.

   $$10x + 5x + 4x = 60$$

Featuring Fractions

Fractions appear frequently in algebraic equations. In the preceding section, I show you how to remove the fractions from an equation when you have the right situation. In this section, I show you how to leave in the fraction, take advantage of the fractional setup, and use it to your advantage.

Promoting practical proportions

A *proportion* is an equation. It consists of two ratios (fractions) set equal to one another. When you write $\frac{6}{12}=\frac{1}{12}$, you're writing a proportion. Before I show you how proportions are solved in algebra problems, I have some properties to share.

Given the proportion $\frac{a}{b} = \frac{c}{d}$:

- ✔ The cross products are equal: $ad = bc$.
- ✔ The reciprocals are equal to one another: $\frac{b}{a} = \frac{d}{c}$.
- ✔ You can reduce the fractions vertically, as usual: $\frac{e \cdot f}{e \cdot g} = \frac{c}{d}$.
- ✔ You can reduce horizontally, across the equal sign: $\frac{e \cdot f}{b} = \frac{e \cdot g}{d}$ or $\frac{a}{e \cdot f} = \frac{c}{e \cdot g}$.

Now I use some of the properties of proportions to solve equations.

Solve for x: $\frac{3x-5}{x+3} = \frac{24}{15}$.

Before cross-multiplying, reduce the fraction on the right by dividing the numerator and denominator by 3:

$$\frac{3x-5}{x+3} = \frac{^8\cancel{24}}{_5 15} = \frac{8}{5}$$

Now, using the cross-multiplying rule:

$$(3x - 5) \cdot 5 = (x + 3) \cdot 8$$
$$15x - 25 = 8x + 24$$

Subtract $8x$ from each side, and add 25 to each side:

$$
\begin{array}{rcr}
15x - 25 &=& 8x + 25 \\
-8x & & -8x \\
\hline
7x - 25 &=& 24 \\
+25 & & +25 \\
\hline
7x &=& 49
\end{array}
$$

Finally, divide each side by 7, and you get $x = 7$.

As another example, solve for y: $\frac{8y-10}{3} - \frac{12y-18}{5} = 0$.

The first thing to do is change the equation to a proportion. Move the second fraction to the right side by adding that fraction to each side of the equation:

$$\frac{8y-10}{3} = \frac{12y-18}{5}$$

Now factor the terms in the two numerators and 'reduce horizontally':

$$\frac{\overset{2}{\cancel{2}}(4y-5)}{3}=\frac{\overset{3}{\cancel{6}}(2y-3)}{5}$$

$$\frac{4y-5}{3}=\frac{3(2y-3)}{5}$$

When reducing proportions, you can divide vertically or horizontally, but you can't reduce the fractions diagonally. The diagonal reductions are done when multiplying fractions and you have a multiplication symbol between, not an equal sign between.

Next, cross-multiply and simplify:

$$(4y - 5) \cdot 5 = 3 \cdot 3(2y - 3)$$

$$20y - 25 = 18y - 27$$

Now, solve the equation by subtracting $18y$ from each side and then adding 25 to each side:

$$
\begin{array}{rcl}
20y - 25 &=& 18y - 27 \\
-18y & & -18y \\
\hline
2y - 25 &=& -27 \\
+25 & & +25 \\
\hline
2y &=& -2
\end{array}
$$

And, finally, dividing each side by 2, you see that $y = -1$.

Transforming fractional equations into proportions

Proportions are very nice to work with because of their unique properties of reducing and changing into non-fractional equations. Many equations involving fractions must be dealt with in that fractional form, but other equations are easily changed into proportions. When possible, you want to take advantage of the situations where transformations can be done.

For example, solve the following equation for x:

$$\frac{x+2}{3} - \frac{5x+1}{6} = \frac{3x-1}{2} + \frac{x-9}{8}$$

You could solve the problem by multiplying each fraction by the least common factor of all the fractions: 24. Another option is to find a common denominator for the two fractions on the left and subtract them, and then find a common denominator for the two fractions on the right and add them. Your result is a proportion:

$$\frac{2(x+2)}{2 \cdot 3} - \frac{5x+1}{6} = \frac{4(3x-1)}{4 \cdot 2} + \frac{x-9}{8}$$

$$\frac{2(x+2)-(5x+1)}{6} = \frac{4(3x-1)+(x-9)}{8}$$

$$\frac{2x+4-5x-1}{6} = \frac{12x-4+x-9}{8}$$

$$\frac{-3x+3}{6} = \frac{13x-13}{8}$$

The proportion can be reduced by dividing by 2 horizontally:

$$\frac{-3x+3}{\cancel{6}_3} = \frac{13x-13}{\cancel{8}_4}$$

Now cross-multiply and simplify the products:

$$(-3x + 3) \cdot 4 = 3 \cdot (13x - 13)$$
$$-12x + 12 = 39x - 39$$

Add $12x$ to each side, and then add 39 to each side:

$$\begin{array}{r} -12x+12 = 39x-39 \\ +12x \qquad\quad +12x \\ \hline 12 = 51x-39 \\ +39 \qquad\quad +39 \\ \hline 51 = 51x \end{array}$$

The last step consists of just dividing each side by 51 to get $1 = x$.

Solving for Variables in Formulas

A formula is an equation that represents a relationship between some structures or quantities or other entities. It's a rule that uses mathematical computations and can be counted on to be accurate each time you use it when applied correctly. The following are some of the more commonly used formulas that contain only variables raised to the first power.

- $A=\frac{1}{2}bh$: The area of a triangle involves base and height.
- $I = Prt$: The interest earned uses principal, rate and time.
- $C = 2\pi r$: Circumference is twice π times the radius.
- $°F = 32° + \frac{9}{5}°C$: Degrees Fahrenheit uses degrees Celsius.
- $P = R - C$: Profit is based on revenue and cost.

When you use a formula to find the indicated variable (the one on the left of the equal sign), you just put the numbers in, and out pops the answer. Sometimes, though, you're looking for one of the other variables in the equation and end up solving for that variable over and over.

For example, let's say that you're helping your mum plan a circular rose garden in your backyard. You find edging on sale and can buy a 20-metre roll of edging, a 36-metre roll, a 40-metre roll or a 48-metre roll. You're going to use every bit of the edging and let the length of the roll dictate how large the garden will be. If you want to know the radius of the garden based on the length of the roll of edging, you use the formula for circumference and solve the following four equations:

$$20 = 2\pi r \qquad 36 = 2\pi r \qquad 40 = 2\pi r \qquad 48 = 2\pi r$$

For more on pi (π) and calculating the perimeter (circumference) and area of circles, see Chapter 15.

Another alternative to solving four different equations is to solve for r in the formula and then put the different roll sizes in to the new formula. Starting with $C = 2\pi r$, you divide each side of the equation by 2π, giving you:

$$r = \frac{C}{2\pi}$$

The computations are much easier if you just divide the length of the roll by 2π.

Now I'll show you some examples of solving for one of the variables in an equation. I won't try to come up with any more gardening or other clever scenarios.

Solve for w in the formula for the perimeter of a rectangle: $P = 2(l + w)$.

First, divide each side of the equation by 2 (instead of distributing the 2 through the terms in the binomial):

$$\frac{P}{2} = l + w$$

Now subtract l from each side. You can write the two terms as a single fraction if you want:

$$\frac{P}{2} - l = w \text{ or } \frac{P - 2l}{2} = w$$

Chapter 11

Taking a Crack at
Quadratic Equations

In This Chapter

▶ Dealing with solutions of special quadratic equations

▶ Factoring quadratic equations for solutions

▶ Applying the greatest common factor and not forgetting a solution

▶ Using the quadratic formula

▶ Making quadratic equations work for you

*Q*uadratic (second-degree) equations are nice to work with because they're manageable. Finding the solution or deciding whether a solution exists is relatively easy — easy, at least, in the world of mathematics.

A quadratic equation is a quadratic expression with an equal sign attached. As with linear equations (covered in Chapter 10), specific methods or processes, given in detail in this chapter, are employed to successfully solve quadratic equations. The most commonly used technique for solving these equations is factoring, but a quick-and-dirty rule can also be used for one of the special types of quadratic equations. I have to warn you, though, that just because someone puts in some numbers and makes up a quadratic equation, that doesn't mean there's necessarily a solution or answer to it. (I show you how to tell if there's no answer in this chapter.)

Quadratic equations are important to algebra and many other sciences. Some quadratic equations say that what goes up must come down. Other equations describe the paths that planets and comets take. In all, quadratic equations are fascinating — and just dandy to work with.

Squaring Up to Quadratics

A quadratic equation contains a variable term with an exponent of 2 and no variable term with a higher power.

A quadratic equation has a general form that goes like this: $ax^2 + bx + c = 0$. The constants a, b and c in the equation are real numbers, and a cannot be equal to 0. (If a were 0, you wouldn't have a quadratic equation anymore.)

If the equation looks familiar, it means that you've read Chapter 9, which talks about factoring and working with quadratic expressions. **Remember:** An expression consists of one or more terms but has no equal sign. Adding an equal sign changes the whole picture: Now you have an equation that says something. The equation forms a true statement if the solutions are put in for the variables.

Here are some examples of quadratic equations and their solutions:

- $4x^2 + 5x - 6 = 0$: In this equation, none of the coefficients is 0. The two solutions are $x = -2$ and $x = \frac{3}{4}$.

- $2x^2 - 18 = 0$: In this equation, the b is equal to 0. The solutions are $x = 3$ and $x = -3$.

- $x^2 + 3x = 0$: In this equation, the c is equal to 0. The solutions are $x = 0$ and $x = -3$.

- $x^2 = 0$: In this equation, both b and c are equal to 0. The equation has only one solution, $x = 0$.

A special feature of quadratic equations is that they can, and often do, have two completely different answers. As you see in the preceding examples, three of the equations have different solutions. The last equation has just one solution, but, technically, you count that solution twice, calling it a *double root*. Some quadratic equations have no solutions if you're only considering real numbers, but get real! We stick to real solutions for now.

How did I find all those solutions in the examples? I used the methods on solving quadratic equations that are found in this chapter. My goal in this section is to get you used to having two different answers that work. You can jump ahead, though, if your curiosity is getting the better of you about other quadratic equations.

How can an equation have two answers? Which answer do you use in an application or story problem? For example, if the story problem asks about how much something costs, how can there be two correct answers? Well, sometimes there are two right answers to the application, but usually one of the answers doesn't really make sense in the particular situation. The nonsensical answer does solve the equation you set up and just comes along as extra baggage. When faced with two answers, you have to make a decision as to whether to pay attention to the extra answer.

Let me show you two examples of problems using quadratic equations that end up with two answers. In the first example, you see that both answers can work. In the second example, only one answer works.

A ball is thrown upward into the air by a person standing on a 16-metre-high wall. The height, h (in metres), of the ball after t seconds is given by the quadratic equation: $h = -16t^2 + 80t + 16$. When is the ball 80 metres in the air?

Don't worry about where I got the equation; it's something discussed in physics and in many maths classes.

I want to figure out when the ball is 80 metres above the ground, so I put in 80 for the height, h:

$$80 = -16t^2 + 80t + 16$$

I just happen to know that when t is equal to 1 or 4, the equation is true. Well, I don't actually know that, but I used the methods from this chapter for finding the solution of a quadratic equation to get the answers. Again, I'm showing you how two answers can work and both make sense.

When $t = 1$:

$$80 = -16(1)^2 + 80(1) + 16 = -16 + 80 + 16 = 80$$

And when $t = 4$:

$$80 = -16(4)^2 + 80(4) + 16 = -256 + 320 + 16 = 80$$

Both work! So this equation says that when t equals 1 (after 1 second) and when t equals 4 (after 4 seconds) the ball is 80 metres in the air. The first time, the ball is going upward, and the second time, the ball is falling toward the ground. If you throw a ball up into the air from 16 metres high, then the ball could go up, pass the 80-metre level, go higher than that, and then be at the 80-metre level again on the way down.

The next example using a quadratic equation has two answers, but only one makes any sense in the actual problem. The answers both work in the equation, but only one answers the question.

You're the controller of Whatchamacallits Company, and you use the following cost function to tell you the total cost of producing n units: $C(n) = 0.04n^2 + 2n + 100$. You get an order from a customer who says he has only \$124 to work with. How many units can the customer buy?

Replace the $C(n)$ with 124 and the equation reads: $124 = 0.04n^2 + 2n + 100$. Solving for n, you get that if $n = 10$ or if $n = -60$, either will make the equation into a true statement: $C(10) = 124$ and $C(-60) = 124$.

Getting these two answers — one of them negative — happens frequently when you use an equation to model what happens in real life. The equation usually works wonderfully to give you answers, but you can't use it beyond what's reasonable.

With this particular equation, it wouldn't make sense to use negative numbers for n because you can't manufacture a negative number of units. And in other situations it also may not be reasonable to use values of n up in the billions or trillions.

The price that's paid for using these nice equations is that they have to be used under reasonable circumstances.

Rooting Out Results from Quadratic Equations

The general quadratic equation has the form $ax^2 + bx + c = 0$, and b or c or both of them can be equal to 0. This section shows you how nice it is — and how easy it is to solve equations — when b is equal to 0.

The first 20 perfect squares (products of a number times itself) are 1, 4, 9, 16, 25, 36, 49, 64, 81, 100, 121, 144, 169, 196, 225, 256, 289, 324, 361 and 400. Notice that the square numbers go from a low of 1 to a high of 400. There aren't any other perfect squares between the ones listed. That means that the other 380 numbers between 1 and 400 are *not* perfect squares. The perfect squares all have nice square roots. The square root of 121 is 11; the square root of 256 is 16. Isn't that nice? But the square root of 200 isn't nice at all; it's an irrational number.

Irrational numbers don't terminate or repeat themselves after the decimal point. For example, the square root of 2, an irrational number, is 1.414213562373 ... An irrational number can't ever be written as a fraction. Irrationals are just as their name describes: Wild and unpredictable.

The roots have decimal values that can be approximated with a calculator, though.

Don't worry if you don't recognise some of the larger squares because they aren't used frequently, and you usually get some sort of a hint that the number is a perfect square when you're doing a problem. Sometimes the hint comes from the wording of the problem — it may talk about a square room or sides of a right-angle triangle. Sometimes the hint is just that it'd be so nice if it were square.

And here's a twist to square roots and squares. Usually, if you're asked for the square root of 25, you say, 'Five.' Well, that's right of course, but that's just the principal square root. When solving quadratic equations, you start with statements involving a variable squared, so you usually have two solutions for the equations. In the case of $x^2 = 25$, the two solutions are $+5$ and -5.

The *principal square root* of a number is just the positive number that, when multiplied by itself, gives you the original number. The principal square root of 49 is 7. When doing a square root to solve an equation, both the principal square root and its inverse (the negative one) are used. So, under certain circumstances, such as solving quadratic equations, you have to consider that other answer, too.

The following is the rule for some special quadratic equations — the ones where $b = 0$. They start out looking like $ax^2 + c = 0$, but the c is usually negative, giving you $ax^2 - c = 0$ and the equation is rewritten as $ax^2 = c$.

If $x^2 = k$, then $x = \pm\sqrt{k}$ or if $ax^2 = c$, then $x = \pm\sqrt{\frac{c}{a}}$. If the square of a variable is equal to the number k, the variable is equal to the principal square root of k or its opposite.

The following examples show you how to use this square-root rule on quadratic equations where $b = 0$.

Solve for x in $x^2 = 49$.

Using the square root rule, $x = \pm\sqrt{49} = \pm 7$. Checking, $(7)^2 = 49$ and $(-7)^2 = 49$.

Solve for m in $3m^2 + 4 = 52$.

This equation isn't quite ready for the square-root rule. Add -4 to each side:

$$3m^2 = 48$$

Now divide each side by 3:

$$m^2 = 16$$

So $m = \pm\sqrt{16} = \pm4$.

Solve for p in $p^2 + 11 = 7$.

Add -11 to each side to get $p^2 = -4$. Oops! What number times itself is equal to -4? The answer is: 'None that you can imagine!'

Mathematicians have created numbers that don't actually exist so that these problems can be finished. The numbers are called *imaginary numbers*, but this section is concerned with the less-heady numbers. So, this problem doesn't have an answer, if you're looking for a real number.

Here's another example. Solve for q in $(q + 3)^2 = 25$.

In this case, you end up with two completely different answers, not one number and its opposite. Use the square-root rule, first, to get $q + 3 = \pm\sqrt{25} = \pm5$.

Now you have two different linear equations to solve:

$$q + 3 = +5 \quad q + 3 = -5$$

Subtracting 3 from each side of each equation, the two answers are:

$$q = 2 \quad q = -8$$

This problem definitely needs to be checked. Putting in the 2:

$$(2 + 3)^2 = 25$$
$$5^2 = 25$$

Putting in the -8:

$$(-8 + 3)^2 = 25$$
$$(-5)^2 = 25$$

Yes, they both work!

Factoring for a Solution

This section is where running through all the factoring methods can really pay off. (Refer to Chapters 8 and 9 for all the details.) In most quadratic equations, factoring is used rather than the square-root rule method covered in the preceding section. The square-root rule is used only when $b = 0$ in the quadratic equation $ax^2 + bx + c = 0$. Factoring is used when $c = 0$ or when neither b nor c is 0.

A very important property used along with the factoring to solve these equations is the multiplication property of zero. This is a very straightforward rule — and it even makes sense. Use the greatest common factor and the multiplication property of zero when solving quadratic equations that aren't in the form for the square-root rule.

Zeroing in on the multiplication property of zero

Before you get into factoring quadratics for solutions, you need to know about the multiplication property of zero. You may say, 'What's there to know? Zero multiplies anything and leaves nothing. It wipes out everything!' True enough, but 0 has another nice property that is the basis of much equation solving in algebra. By itself, 0 is nothing. Put it as the result of a multiplication problem, and you really have something: The *multiplication property of zero*.

The *multiplication property of zero* (MPZ) states that if $p \times q = 0$, then either $p = 0$ or $q = 0$. At least one of them must be equal to 0. Note that this can also be referred to as the Null Factor Law (NFL).

This may seem obvious, but think about it. No other number has such a power over all other numbers. If you say that $p \times q = 12$, you can't predict a thing about p or q alone. These variables could be any number at all — positive, negative, fractional, radical, or a mixture of these. A product of 0, however, leads to one conclusion: One of the multipliers must be 0. No other means of arriving at a 0 product exists. Why is this such a big deal? Let me show you a few equations and how the NFL works.

Find the value of x if $3x = 0$.

$x = 0$ because 3 can't be 0. Using the NFL, if the one factor isn't 0, then the other must be 0.

Getting the quadratic second-degree

The word *quadratic* is used to describe equations that have a second-degree term. Why, then, is the prefix *quad-*, which means 'four', used in a second-degree equation? It appears that this came about because a square is the regular four-sided figure, whose sides are the same. The area of a square with sides x long would be x^2. So 'squaring' in this case is raising to the second power.

Find the value of x and y if $xy = 0$.

You have two possibilities in this equation. If $x = 0$, then y can be any number, even 0. If $x \neq 0$, then y must be 0, according to the NFL.

Solve for x in $x(x - 5) = 0$.

Again, you have two possibilities. If $x = 0$, the product of $0(-5) = 0$. The other choice is when $x = 5$. Then you have $5(0) = 0$.

Assigning the greatest common factor and multiplication property of zero to solving quadratics

Factoring is relatively simple with only two terms that have a common factor. This is true in quadratic equations of the form

$ax^2 + bx = 0$ (where $c = 0$).

The two terms left have the common factor of x, at least. You find the greatest common factor (GCF) and factor that out, and then use the NFL to solve the equation.

The following examples make use of the fact that the constant term is 0, and the two terms have a common factor of at least an x.

Use factoring to solve for x in $x^2 - 7x = 0$.

The GCF of the two terms is x, so write the left side in factored form:

$x(x - 7) = 0$

Use the NFL to say that either $x = 0$ or $x - 7 = 0$. The first equation gives you $x = 0$, and the second solves to give you $x = 7$.

Solve for x in $6x^2 + 18x = 0$.

The GCF of the two terms is $6x$, so write the left side in factored form:

$$6x(x + 3) = 0$$

Use the NFL to say that $6x = 0$ or $x + 3 = 0$, which gives you the two solutions $x = 0$ or $x = -3$.

Technically, I could have written three different equations from the factored form:

$$6 = 0 \qquad x = 0 \qquad x + 3 = 0$$

The first equation, $6 = 0$, makes no sense — it's an impossible statement. So you either ignore setting the constants equal to 0 or combine them with the factored-out variable, where they'll do no harm.

Because $c = 0$ in so many quadratic equations, it might be useful to have a rule or formula for what the solutions are every time. So, to create a rule, solve for x in this general quadratic equation where $c = 0$: $ax^2 + bx = 0$.

The GCF of the two terms is x, so write the left side in factored form:

$$x(ax + b) = 0$$

Use the NFL to say that $x = 0$ or $ax + b = 0$. The first part of this is pretty clear. And this $x = 0$ business seems to crop up every time. The second part takes careful solving of the linear equation. Subtract b from each side:

$$ax + b - b = 0 - b$$

Divide each side by a:

$$\frac{ax}{a} = \frac{-b}{a}$$
$$x = -\frac{b}{a}$$

So the two solutions of $ax^2 + bx = 0$ are $x = 0$ and $x = \frac{-b}{a}$.

This recognisable pattern can help you solve these types of equations. You can use this as a formula and not have to do the factoring and solving each time.

Missing the $x = 0$, a full half of the solution, is an amazingly frequent occurrence. You don't notice the lonely little x in the front of the parentheses and forget that it gives you one of the two answers. Be careful.

Solving Quadratics with Three Terms

Quadratic equations are basic not only to algebra but also to physics, business, astronomy and many other applications. By solving a quadratic equation, you get answers to questions such as, 'When will the rock hit the ground?' or 'When will the profit be greater than 100 per cent?' or 'When, during the year, will the earth be closest to the sun?'

In the two previous sections, either b or c has been equal to 0 in the quadratic equation $ax^2 + bx + c = 0$. Now I won't let anyone skip out. In this section, each of the letters, a, b and c is a number that is not 0.

To solve a quadratic equation, moving everything to one side with 0 on the other side of the equal sign is the most efficient method. Factor the equation if possible, and use the NFL after you factor. If you don't have three terms in the equation, refer to the previous sections.

In the following example, I list the steps you use for solving a quadratic trinomial by factoring.

Solve for x in $x^2 - 3x = 28$. Follow these steps:

1. **Move all the terms to one side. Get 0 alone on the right side.**

   In this case, you can subtract 28 from each side:

   $$x^2 - 3x - 28 = 0$$

   The standard form for a quadratic equation is $ax^2 + bx + c = 0$. (Refer to Chapter 9 for more on quadratic equations and how to factorise.)

2. **Determine all the ways you can multiply two numbers to get a.**

   In $x^2 - 3x - 28 = 0$, $a = 1$, which can only be 1 times itself.

3. **Determine all the ways you can multiply two numbers to get c (ignore the sign for now).**

   28 can be 1×28, 2×14, or 4×7.

4. **Factor.**

   If c is positive, find an operation from your Step 2 list and an operation from your Step 3 list that match so that the sum of their cross-products is the same as b.

   If c is negative, find an operation from your Step 2 list and an operation from your Step 3 list that match so that the difference of their cross-products is the same as b.

   In this problem, c is negative, and the difference of 4 and 7 is 3. Factoring, you get $(x - 7)(x + 4) = 0$.

5. **Use the NFL.**

   Either $x - 7 = 0$ or $x + 4 = 0$; now try solving for x by getting x alone to one side of the equal sign.

   - $x - 7 + 7 = 0 + 7$ gives you that $x = 7$.

   - $x + 4 - 4 = 0 - 4$ gives you that $x = -4$.

   So the two solutions are $x = 7$ or $x = -4$.

6. **Check your answer.**

   If $x = 7$, then $(7)^2 - 3(7) = 49 - 21 = 28$.

   If $x = -4$, then $(-4)^2 - 3(-4) = 16 + 12 = 28$.

   They both check.

Factoring to solve quadratics sounds pretty simple on the surface. But factoring *trinomial equations* — those with three terms — can be a bit less simple. If a quadratic with three terms can be factored, the product of two binomials is that trinomial. If the quadratic equation with three terms can't be factored, then use the quadratic formula (see 'Figuring Out the Quadratic Formula', later in this chapter).

The product of the two binomials $(ax + b)(cx + d)$ is equal to the trinomial $acx^2 + (ad + bc)x + bd$. This is a fancy way of showing what you get from using FOIL when multiplying the two binomials together.

Now, on to using unFOIL. If you need more of a review of FOIL and unFOIL, check out Chapter 9.

The following examples all show how factoring and the NFL allow you to find the solutions of a quadratic equation with all three terms showing.

Solve for x in $x^2 - 5x - 6 = 0$.

1. **The equation is in standard form, so you can proceed.**

2. **Determine all the ways you can multiply to get a.**

   $a = 1$, which can only be 1 times itself. If there are two binomials that the left side factors into, they must each start with an x because the coefficient of the first term is 1.

   $(x)(x) = 0$

3. **Determine all the ways you can multiply to get c.**

   $c = -6$, so, looking at just the positive factors, you have 1×6 or 2×3.

4. **Factor.**

   To decide which combination should be used, look at the sign of the last term in the trinomial, the 6, which is negative. This tells you that you have to use the *difference* of the absolute value of the two numbers in the list (think of the numbers without their signs) to get the middle term in the trinomial, the -5. In this case, one of the 1 and 6 combinations work, because their difference is 5. If you use the $+1$ and -6, then you get the -5 immediately from the cross-product in the FOIL process. So $(x - 6)(x + 1) = 0$.

5. **Use the NFL.**

   Using the NFL, $x - 6 = 0$ or $x + 1 = 0$. This tells you that $x = 6$ or $x = -1$.

6. **Check.**

   If $x = 6$, then $(6)^2 - 5(6) - 6 = 36 - 30 - 6 = 0$.

   If $x = -1$, then $(-1)^2 - 5(-1) - 6 = 1 + 5 - 6 = 0$.

   They both work!

Here's another example. Solve for x in $6x^2 + x = 12$.

1. **Put the equation in the standard form.**

   The first thing to do is to add -12 to each side to get the equation into the standard form for factoring and solving:

   $6x^2 + x - 12 = 0$

   This one will be a bit more complicated to factor because the 6 in the front has a couple of choices of factors, and the 12 at the end also has several choices. The trick is to pick the correct combination of choices.

2. **Find all the combinations that can be multiplied to get a.**

   You can get 6 with 1×6 or 2×3.

3. Find all the combinations that can be multiplied to get *c*.

You can get 12 with 1×12, 2×6, or 3×4.

4. Factor.

You have to choose the factors to use so that the difference of their cross-products (outer and inner) is 1, the coefficient of the middle term. How do you know this? Because the 12 is negative, in this standard form, and the value multiplying the middle term is assumed to be 1 when there's nothing showing.

Looking this over, you can see that using the 2 and 3 from the 6 and the 3 and 4 from the 12 will work: $2 \times 4 = 8$ and $3 \times 3 = 9$. The difference between the 8 and the 9 is, of course, 1. You can worry about the sign later.

Fill in the binomials and line up the factors so that the 2 multiplies the 4 and the 3 multiplies the 3, and you get a 6 in the front and 12 at the end. Whew!

$$(2x\ 3)(3x\ 4) = 0$$

The quadratic has a + on the term in the middle, so I need the bigger product of the outer and inner to be positive. I get this by making the $9x$ positive, which happens when the 3 is positive and the 4 is negative.

$$(2x + 3)(3x - 4) = 0$$

5. Use the NFL to solve the equation.

The trinomial has been factored. The NFL tells you that either $2x + 3 = 0$ or $3x - 4 = 0$. If $2x + 3 = 0$, then $2x = -3$ or $x = -\frac{3}{2}$. If $3x - 4 = 0$, then $3x = 4$ or $x = \frac{4}{3}$.

6. Check your work.

When $x = \frac{3}{2}$, then $6\left(-\frac{3}{2}\right)^2 + \left(-\frac{3}{2}\right) = 12$ and $6\left(\frac{9}{4}\right) - \frac{3}{2} = \frac{27}{2} - \frac{3}{2} = \frac{24}{2} = 12$.

When $x = \frac{4}{3}$, then $6\left(\frac{4}{3}\right)^2 + \left(\frac{4}{3}\right) = 12$ and $6\left(\frac{19}{9}\right) + \frac{4}{3} = \frac{32}{3} + \frac{4}{3} = \frac{36}{3} = 12$.

This checking wasn't nearly as fun as some, but it sure does show how well this factoring business can work.

Solve for y in $9y^2 - 12y + 4 = 0$.

1. This is already in the standard form.

2. Find all the numbers that multiply to get *a*.

The factors for the 9 are 1×9 or 3×3.

3. Find all the numbers that multiply to get *c*.

The factors for c are 1×4 or 2×2.

4. **Factor.**

   Using the 3s and the 2s is what works because both cross-products are 6, and you need a sum of 12 in the middle. So,

   $$9y^2 - 12y + 4 = (3y - 2)(3y - 2) = 0$$

   Notice that I put the negative signs in because the 12 needs to be a negative sum.

5. **Use the NFL to solve the equation.**

   The two factors are the same here. That means that using the NFL gives you the same answer twice. When $3y - 2 = 0$, solve this for y. First add the 2 to each side, and then divide by 3. The solution is $y = \frac{2}{3}$. This is a double root, which, technically, has only one solution, but it occurs twice.

A double root occurs in quadratic trinomial equations that come from perfect-square binomials. Perfect-square binomials are discussed in Chapter 7, if you need a refresher. These perfect-square binomials are no more than the result of multiplying a binomial times itself. That's why, when they're factored, there's only one answer — it's the same one for each binomial.

Here's another example. Solve for z in $12z^2 - 4z - 8 = 0$.

1. **This quadratic is already in standard form.**

   You can start out by looking for combinations of factors for the 12 and the 8, but you may notice that all three terms are divisible by 4. To make things easier, take out that GCF first, and then work with the smaller numbers in the parentheses.

   $$12x^2 - 4z - 8 = 4(3z^2 - z - 2) = 0$$

2. **Find the numbers that multiply to get 3.**

   $$3 = 1 \times 3$$

3. **Find the numbers to multiply to get 2.**

   $$2 = 1 \times 2$$

4. **Factor.**

   This is really wonderful, especially because the 3 and 2 are both prime and can be factored only one way. Your only chore is to line up the factors to ensure a difference of 1 between the cross-products.

   $$4(3z^2 - z - 2) = 4(3z\ 2)(z\ 1) = 0$$

Because the middle term is negative, you need to make the larger
product negative, so put the negative sign on the 1.

$$4(3z + 2)(z - 1) = 0$$

5. Use the NFL to solve for the value of z.

This time, when you use the NFL, you need to consider three factors.
Either $4 = 0$, $3z + 2 = 0$, or $z - 1 = 0$. The first equation is impossible;
4 doesn't ever equal 0. But the other two equations give you answers. If

$$3z + 2 = 0, \text{ then } z = -\tfrac{2}{3}. \text{ If } z - 1 = 0, \text{ then } z = 1.$$

6. Check.

If $z = -\tfrac{2}{3}$, then $12\left(-\tfrac{2}{3}\right)^2 - 4\left(-\tfrac{2}{3}\right) - 8 = 0$ and $12\left(\tfrac{4}{9}\right) + \tfrac{8}{3} - 8 = \tfrac{16}{3} + \tfrac{8}{3} - 8 = \tfrac{24}{3} - 8 = 8 - 8 = 0.$

If $z = 1$, then $12(1)^2 - 4(1) - 8 = 12 - 4 - 8 = 0.$

When checking your solution(s) — always use the original equation (the
version before you did anything to it).

Applying Quadratic Equation Solutions

Quadratic equations are found in many mathematics, science, and business
applications; that's why they're studied so much. The graphs of quadratic
equations are always U-shaped, with an extreme point that's highest, lowest,
farthest left or farthest right. That extreme point is often the answer to
a question about the situation being modelled by the quadratic. In other
applications, you want the point(s) at which the U-shaped curve crosses
an axis; those points are found by finding solutions to setting the quadratic
equal to 0. In this section, I show you some examples of how quadratic
equations are used in applications.

In physics, an equation that tells you how high an object is after a certain
amount of time can be written $h = -16t^2 + v_0 t + h_0$. In this equation, the $-16t^2$
part accounts for the pull of gravity on the object. The number representing
v_0 is the initial velocity — what the speed is at the very beginning. The h_0 is
the starting height — the height in metres of the building, cliff or stool from
which the object is thrown or shot or dropped. The variable t represents
time — how many seconds or minutes have passed.

For example, say a stone was thrown upward from the top of a 40-metre
building with a beginning speed of 128 metres per second. When was the
stone 296 metres up in the air?

Replacing the height, h, with the 296, the v_0 with 128, and the h_0 with 40, the equation now reads: $296 = -16t^2 + 128t + 40$. You can solve it using the following steps:

1. **Put the equation in standard form.**

   Add -296 to each side.

   $$0 = -16t^2 + 128t - 256$$

2. **Factor out the GCF.**

   In this case, the GCF is -16.

   $$0 = -16(t^2 - 8t + 16)$$

3. **Factor the quadratic trinomial inside the parentheses.**

   $$0 = -16(t - 4)^2$$

4. **Use the NFL to solve for the variable.**

   $$t - 4 = 0, \ t = 4$$

   After 4 seconds, the stone will be 296 metres up in the air.

The next example gets into the business side of these equations. The profit earned from producing and selling items is determined by subtracting the cost from the revenue. Equations can act as models for the amount of profit based on the number of items produced and sold. This example shows you how a model works.

The profit from manufacturing and selling Flimsy Flippers is determined using $P(f) = -0.1f^2 + 22f - 210$, where f is the number of pairs of flippers. How many pairs of flippers must be produced and sold to have a positive profit?

The graph of the profit function is a *parabola* (a U-shaped curve) opening downward. (For more on these graphs, refer to Chapter 14.) What you need to find is when the profit goes from negative to positive and then back down to negative. (Profit decreases with too much overtime or outsourcing from too many items being produced.) The function changes from negative to positive and positive to negative when $P(f) = 0$. So the answer is found by solving the quadratic equation $-0.1f^2 + 22f - 210 = 0$.

The first thing to do is to factor -0.1 out of each term. It's hard to unFOIL quadratics when the lead coefficient is negative and even harder when it's a decimal or fraction. Factoring out a GCF of -0.1, you get

$$-0.1(f^2 - 220f + 2{,}100) = 0$$

To factor the quadratic trinomial, you need to find two factors of 2,100 with a sum of 220. The two factors are 210 and 10. Factoring the quadratic, you get

$$-0.1(f - 10)(f - 210)$$

Using the NFL, you get that the profit is 0 when $f = 10$ or when $f = 210$. And I show you that the profit is positive for numbers between 10 and 210 (and negative otherwise). Here are some of the function values:

$$f(5) = -0.1(5)^2 + 22(5) - 210 = -2.5 + 110 - 210 = -102.5$$

$$f(15) = -0.1(15)^2 + 22(15) - 210 = -22.5 + 330 - 210 = 97.5$$

$$f(100) = -0.1(100)^2 + 22(100) - 210 = -1{,}000 + 2{,}200 - 210 = 990$$

$$f(205) = -0.1(205)^2 + 22(205) - 210 = -4{,}202.5 + 4{,}510 - 210 = 97.5$$

$$f(250) = -0.1(250)^2 + 22(250) - 210 = -6{,}250 + 5{,}500 - 210 = -960$$

Figuring Out the Quadratic Formula

The quadratic formula is special to quadratic equations. A quadratic equation, $ax^2 + bx + c = 0$, can have as many as two solutions, but there may be only one solution or even no solution at all.

a, b and c are any real numbers. The a can't equal 0, but the b or c can equal 0.

The quadratic formula allows you to find solutions when the equations aren't very nice. Numbers aren't *nice* when they're funky fractions, indecent decimals with no end, or raucous radicals.

The quadratic formula says that if an equation is in the form $ax^2 + bx + c = 0$, its solutions, the values of x, can be found with the following:

$$x = \frac{-b \pm \sqrt{b^2 - 4ac}}{2a}$$

You see an operation symbol, $\pm$, in the formula. The symbol is shorthand for saying that the equation can be broken into two separate equations, one using the plus sign and the other using the minus sign. They look like the following:

$$x = \frac{-b + \sqrt{b^2 - 4ac}}{2a}$$

$$x = \frac{-b - \sqrt{b^2 - 4ac}}{2a}$$

Can you see the difference between the two equations? The only difference is the change from the plus sign to the minus sign before the radical.

You can apply this formula to *any* quadratic equation to find the solutions — whether it factors or not. Let me show you some examples of how the formula works.

Use the quadratic formula to solve $2x^2 + 7x - 4 = 0$.

Refer to the standard form of a quadratic equation where the coefficient of x^2 is a, the coefficient of x is b and the constant is c. In this case, $a = 2$, $b = 7$ and $c = -4$. Inserting those numbers into the formula, you get

$$x = \frac{-7 \pm \sqrt{7^2 - 4(2)(-4)}}{2(2)}$$

Now, simplifying, and paying close attention to the order of operations, you get

$$x = \frac{-7 \pm \sqrt{49 - (-32)}}{4} = \frac{-7 \pm \sqrt{81}}{4} = \frac{-7 \pm 9}{4}$$

The two solutions are found by applying the + in front of the 9 and then the − in front of the 9.

$$x = \frac{-7 + 9}{4} = \frac{2}{4} = \frac{1}{2}$$
$$x = \frac{-7 - 9}{4} = \frac{-16}{4} = -4$$

Whenever the answers you get from using the quadratic formula come out as integers or fractions, it means that the trinomial could have been factored. It doesn't mean, though, that you shouldn't use the quadratic formula on factorable problems. Sometimes it's easier to use the formula if the equation has really large or nasty numbers. In general, though, it's quicker to factor using unFOIL and then the NFL when you can. Just to illustrate this, look at the previous example when it's solved using factoring and the NFL:

$$2x^2 + 7x - 4 = (2x - 1)(x + 4) = 0$$

Then using the NFL, you get $2x - 1 = 0$ or $x + 4 = 0$, so $x = \frac{1}{2}$ or $x = -4$.

So, what do the results look like when the equation can't be factored? The next example shows you.

Here are two things to watch out for when using the quadratic formula:

- ✔ **Don't forget that −b means to use the *opposite* of b.** If the coefficient *b* in the standard form of the equation is a positive number, change it to a negative number before inserting into the formula. If *b* is negative, change it to positive in the formula.

- ✔ **Be careful when simplifying under the radical.** The order of operations dictates that you square the value of *b* first, and then multiply the last three factors together before subtracting them from the square of *b*. Some sign errors can occur if you're not careful.

Solve for x using the quadratic formula in $2x^2 + 8x + 7 = 0$.

In this problem, you let $a = 2$, $b = 8$ and $c = 7$ when using the formula:

$$x = \frac{-8 \pm \sqrt{8^2 - 4(2)(7)}}{2(2)} = \frac{-8 \pm \sqrt{64 - 56}}{4} = \frac{-8 \pm \sqrt{8}}{4}$$

The radical can be simplified because $\sqrt{8} = \sqrt{4} \cdot \sqrt{2} = 2\sqrt{2}$, so

$$x = \frac{-8 \pm 2\sqrt{2}}{4} = \frac{-^4 8 \pm 2\sqrt{2}}{^2 4} = \frac{-4 \pm \sqrt{2}}{2}$$

Be careful when simplifying this expression: $\frac{\left(-4 + \sqrt{2}\right)}{2} \neq -2 + \sqrt{2}$. Both terms in the numerator of the fraction have to be divided by the 2.

Here are the decimal equivalents of the answers:

$$\frac{-4 + \sqrt{2}}{2} \approx \frac{-4 + 1.414}{2} = \frac{-2.586}{2} = -1.293$$

$$\frac{-4 - \sqrt{2}}{2} \approx \frac{-4 - 1.414}{2} = \frac{-5.414}{2} = -2.707$$

When you check these answers, what do the estimates do? If $x = -1.293$, then $2(-1.293)^2 + 8(-1.293) + 7 = 3.343698 - 10.344 + 7 = -0.000302$.

That isn't 0! What happened? Is the answer wrong? No, it's okay. The rounding caused the error — it didn't come out exactly right. This happens when you use a rounded value for the answer, rather than the exact radical form. An estimate was used for the answer because the square root of a number that is not a perfect square is an irrational number, and the decimal never ends. Rounding the decimal value correct to three decimal places seemed like enough decimal places.

You shouldn't expect the check to come out to be *exactly* 0. In general, if you round the number you get from your check to the same number of places that you rounded your estimate of the radical, you should get the 0 you're aiming for.

In some instances, questions will ask you to leave the answer in *surd form* or in exact form. This simply means leave your answer under the root sign. In other instances, the question will ask you to find the decimal equivalents, so make sure you read the question very carefully!

Here's a final example. Use the quadratic formula to solve $2x^2 + 7x - 5 = 0$, and give your answer in exact (or surd) form.

In this quadratic, $a = 2$, $b = 7$, and $c = -5$.

Here's the workings for putting the numbers into the formula.

$$x = \frac{-b \pm \sqrt{b^2 - 4ac}}{2a}$$

$$= \frac{-7 \pm \sqrt{7^2 - 4 \times 2 \times -5}}{2 \times 2}$$

$$= \frac{-7 \pm \sqrt{49 - -40}}{4}$$

$$= \frac{-7 \pm \sqrt{49 + 40}}{4}$$

$$= \frac{-7 \pm \sqrt{89}}{4}$$

$$x = \frac{-7 + \sqrt{89}}{4} \text{ or } x = \frac{-7 - \sqrt{89}}{4}$$

Generally, mathematicians (and so your maths teachers) don't like a negative sign to come first, so chances are that these answers would be rewritten as the following.

$$x = \frac{\sqrt{89} - 7}{4} \text{ or } x = \frac{-\sqrt{89} - 7}{4}$$

Now, if the question asked you to give an approximate answer, correct to two decimal places, you need to evaluate each possible value of x.

$$x = \frac{-7 + \sqrt{89}}{4}$$

$$= 0.608\,495\ldots$$

$$\approx 0.61 \,(\text{correct to two decimal places})$$

or

$$x = \frac{-7 - \sqrt{89}}{4}$$

$$= -4.108\,495\ldots$$

$$\approx -4.11 \,(\text{correct to two decimal places})$$

Most calculators have a key that will convert between exact answers and approximate answers. On my calculator, this key looks something like this $<\,>$. If you can't find the correct key on your calculator, ask your teacher to show you where it is! This is such a great key to find because it enables you to type in a fraction and convert it to a decimal, convert a decimal to a fraction and also, as needed in the preceding example, convert an exact answer to an approximate answer.

Part IV
Applying Algebra and Understanding Geometry

Top Five Elements of a Cartesian Plane

✔ Two *lines* cross one another at right angles to form four sections or quadrants.

✔ The two lines, or *axes* (pronounced *ax*-eez), are number lines usually marked with the integers (positive and negative whole numbers and 0). The positives go upward on the vertical axis and to the right on the horizontal axis.

✔ The line going left and right — the horizontal line — is the x-axis; the line going up and down — the vertical line — is the y-axis.

✔ The little marks on the axes are called *tick marks*. They're all uniformly spaced and are usually labelled with the integers, negative to positive, left to right, and downward to upward, with 0 in the middle, at the point where the axes meet.

✔ The four *quadrants* are numbered I, II, III and IV, with capital roman numerals starting with the upper-right quadrant and going anticlockwise.

In this part . . .

- ✔ Work with the Cartesian plane and understand graphing basics.

- ✔ Investigate intersecting lines and lines that never meet, and slide down slopes and around circles.

- ✔ Get the measure of perimeters, area and volume.

- ✔ Find out everything you need to know about geometry and those tricky triangles.

Chapter 12

Graphing Basics

In This Chapter

▶ Understanding the Cartesian plane

▶ Pointing at points and calling them by name

▶ Graphing formulas and equations

A picture is worth a thousand words. This saying is especially true in algebra. Pictures or graphs give you an instant impression of what's happening in a situation or what an equation is representing in space. A graph is a drawing that illustrates an algebraic operation, equation or formula in a two-dimensional plane (like a piece of graph paper). A graph allows you to see the characteristics of an algebraic statement immediately, compared to the many words needed to describe what you see in a graph.

Most people are familiar with bar graphs and their rectangles standing on end, which often depict test scores. Pie graphs (circles with wedges of varying sizes) are good for showing relationships between pieces of a whole, such as how money is spent, and line graphs are great for showing the ups and downs of the stock market or how your weight is changing over time.

The graphs in algebra are unique because they reveal relationships that you can use to model a situation: A line can model the depreciation of the value of a car; parabolas can model daily temperature; and a flat, S-shaped curve can model the number of people infected with the flu. All these and other models are useful for illustrating what's happening and predicting what can happen in the future.

Algebraic equations match up with their graphs. With algebraic operations and techniques applied to equations to make them more usable, the equations can be used to predict, project and figure out various problems.

The Cartesian Plane

If you need a quick refresher about how the *x-y* coordinate system, otherwise known as the *Cartesian plane*, works, no worries. Check out Figure 12-1.

In the Cartesian plane, two lines cross one another at right angles to form four sections or *quadrants*. The two lines, or *axes* (pronounced *ax*-eez), are number lines usually marked with the integers (positive and negative whole numbers and 0). The positives go upward on the vertical axis and to the right on the horizontal axis. The line going left and right — the horizontal line — is the *x*-axis; the line going up and down — the vertical line — is the *y*-axis.

The little marks on the axes are called *tick marks*. They're all uniformly spaced (like the tick-tocks of a clock are the same time apart) and are usually labelled with the integers, negative to positive, left to right, and downward to upward, with 0 in the middle, at the point where the axes meet.

Here are some other points to note about the Cartesian plane:

- Points are located within the coordinate plane with pairs of coordinates called *ordered pairs* — like (8, 6) or (−10, 3). The first number, the *x-coordinate*, tells you how far you go right or left; the second number, the *y-coordinate*, tells you how far you go up or down.

- Going anticlockwise from the upper-right section of the Cartesian plane are *quadrants* I, II, III, IV.

- The Pythagorean theorem (see Chapter 15) comes up a lot when you're using the coordinate system because when you go right and then up to plot a point (or left and then down, and so on), you're tracing along the legs of a right-angled triangle; the segment connecting the *origin* (0, 0) to the point then becomes the hypotenuse of the right-angled triangle. In Figure 12-1, you can see the 6-8-10 right-angled triangle in quadrant I.

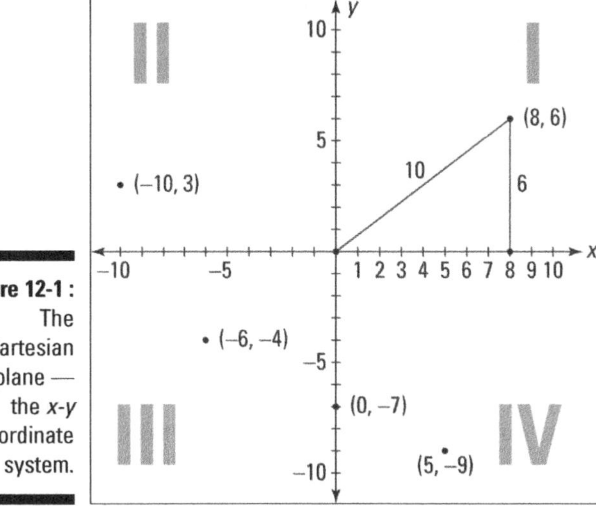

Figure 12-1 :
The Cartesian plane — the *x-y* coordinate system.

Grappling with Graphs

The cartoons in the newspaper often show a worried business person pointing to a graph full of ups and downs — usually the punch line involves a huge drop in sales. As entertaining as these cartoons may be, they also cut right to a major usefulness of graphs: Graphs give instant recognition to the lowest value and the highest value. They give information on trends, patterns and current status. And you can compare two graphs in the same picture.

In almost all cases, a graph of a function or equation in algebra is drawn on a Cartesian plane (refer to the earlier section 'The Cartesian Plane').

Making a point

You do the type of point-finding needed to do graphing when you find the whereabouts of Ballarat, Victoria, at G7 on a road atlas. You move your finger so it's down from the G and across from the 7. Graphing in algebra is just a bit different because numbers replace the letters, and you start in the middle at a point called the *origin*.

Points are dots on a piece of paper or blackboard that represent positions or places with respect to the axes — vertical and horizontal lines — of a graph. The coordinates of a point tell you its exact position on the graph (unlike maps, where G7 can be a big area and you have to look around for the city).

The axes of an algebraic graph are usually labelled with integers, but they can be labelled with any rational numbers, as long as the numbers are the same distance apart from each other, such as the one-quarter distance between $\frac{1}{4}$, $\frac{1}{2}$, and $\frac{3}{4}$.

Ordering pairs, or coordinating coordinates

To actually plot a point in a graph, you need information on where to put that point. That's where ordered pairs come in.

An *ordered pair* is a set of two numbers called *coordinates* that are written inside parentheses with a comma separating them. Some examples are (2, 3), (–1, 4) and (5, 0). When using particular notation, the order matters: The first number, or *x*-coordinate, tells you the point's position with respect to the *x*-axis — how far to the left or right from the origin — and the second number, or *y*-coordinate, tells you the point's position with respect to the *y*-axis — how far up or down from the origin.

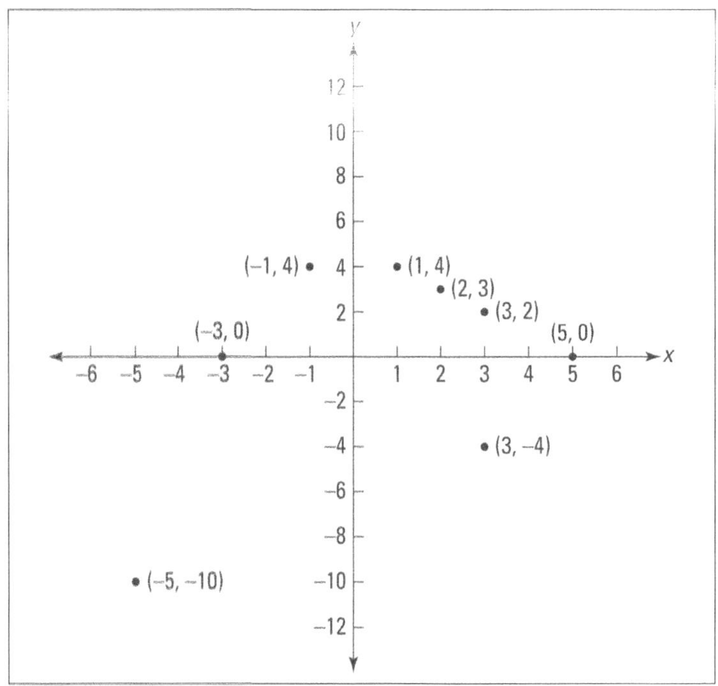

Figure 12-2:
Coordinates and their points on a graph.

For example, the point for the ordered pair (3, 2) is 3 units to the right of the origin and 2 units up from there. Look at Figure 12-2 to see where the points are for several ordered pairs.

Everything starts at the origin — the intersection of the two axes. The ordered pair for the origin is (0, 0). The numbers in this ordered pair tell you that the point didn't go left, right, up or down. Its position is at the starting place.

Notice that point (2, 0) lies right on the *x*-axis. Whenever 0 is a coordinate within the ordered pair, the point must be located on an axis.

Table 12-1 gives you the names of the quadrants, their positions in the Cartesian plane, and the characteristics of coordinate points in the various quadrants. Table 12-2 describes what's happening on the axes as they radiate out from the origin.

Table 12-1	Quadrants		
Quadrant	*Position*	*Coordinate Signs*	*How to Plot*
Quadrant I	Upper-right side	(positive, positive)	Move right and up
Quadrant II	Upper-left side	(negative, positive)	Move left and up
Quadrant III	Lower-left side	(negative, negative)	Move left and down
Quadrant IV	Lower-right side	(positive, negative)	Move right and down

Table 12-2	Axes	
Position	*Coordinate Signs*	*How to Plot*
Right axis	(positive, 0)	Move right and sit on the *x*-axis
Left axis	(negative, 0)	Move left and sit on the *x*-axis.
Upper axis	(0, positive)	Move up and sit on the *y*-axis.
Lower axis	(0, negative)	Move down and sit on the *y*-axis.

Actually Graphing Points

To plot a point, look at the coordinates — the numbers in the parentheses. The first number tells you which way to move, horizontally, from the origin. Place your pencil on the origin and move right if the first number is positive; move left if the first number is negative. Next, from that position, move your pencil up or down — up if the second number is positive and down if it's negative.

The following points are graphed in Figure 12-3. The letters serve as names of the points so you can compare their coordinates.

A (9, 0) B (7, 4)

C (3, 8) D (0, 7)

E (–2, 2) F (–8, 0)

G (–5, –3) H (0, –3)

J (3, –2) K (8, –7)

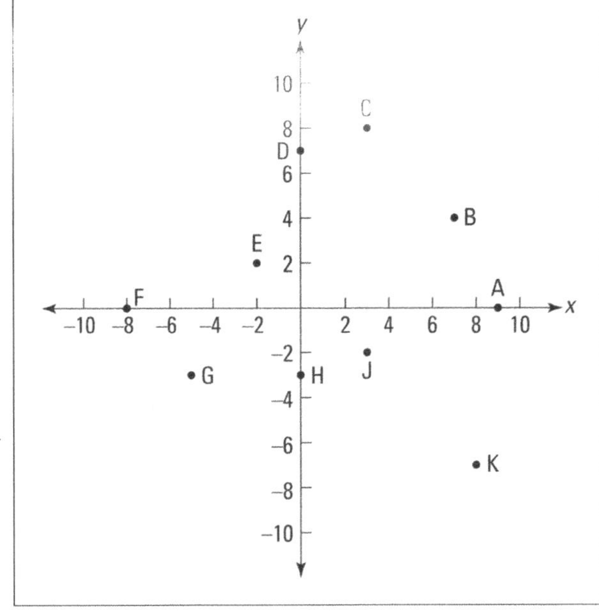

Figure 12-3: Points A through K graphed in the Cartesian plane.

If you have an eagle eye, you may have noticed that I skip from H to J in Figure 12-3. When labelling points on a graph, try to avoid using the letters I and O — these letters are easily mistaken for 1 and 0.

Graphing Is Good

Consider the three ways of expressing the same thing in each of the following examples:

- ✔ **In words:** All the pairs of numbers that add up to 10
- ✔ **In an algebraic equation:** $x + y = 10$
- ✔ **In a graph:** See Figure 12-4.

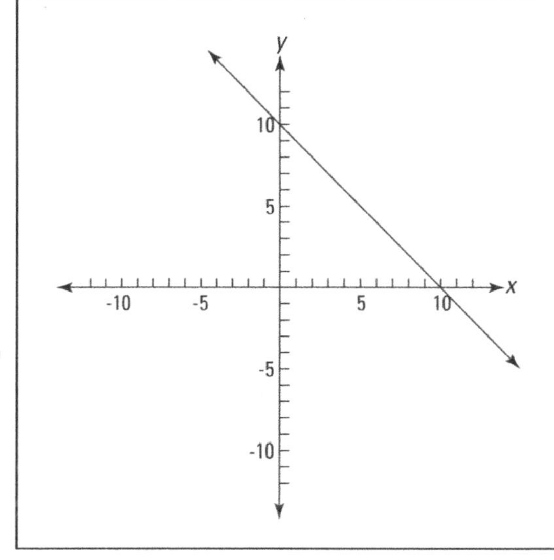

Figure 12-4:
All the possibilities for $x + y = 10$.

And, again:

- ✔ **In words:** All the pairs of numbers you get when you choose the first number and then get the second number by subtracting the first number from its square
- ✔ **In an algebraic equation:** $y = x^2 - x$
- ✔ **In a graph:** See Figure 12-5.

The algebraic equation describes the situation in a more concise manner than the wordy description. The graph, however, gives you a better idea of what's being described than the words or the equation.

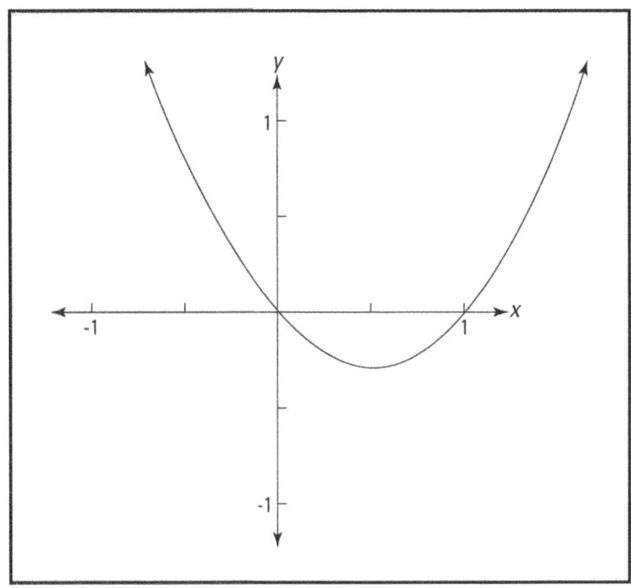

Figure 12-5:
All the possibilities for $y = x^2 - x$.

Graphing Formulas and Equations

An algebraic graph is a picture of the relationship between the two numbers forming the coordinates of a point. The relationship between the coordinates may come in the form of a simple equation such as: $y = x + 3$, which says that whatever the x coordinate is, the y coordinate is 3 bigger. Another relationship or equation might state that the sum of the squares of the two coordinates has to be exactly 25: $x^2 + y^2 = 25$. The relationships are many and varied. I show you several examples of the graphs of the equations or formulas in this section.

Lining up a linear equation

The graph of a linear equation in two variables is a line. For example, the graph of the linear equation $y = x + 3$ is a line that appears to move upward as the x-coordinates increase. I talk more about graphing lines in Chapter 13. For now, I just show you how to do a basic graph.

The graph of $y = x + 3$ goes through all the points in the Cartesian plane that make the equation a true statement. For example, if $x = 2$, then $y = 2 + 3 = 5$, and you have the point (2, 5). Here are some of the points that make the equation true:

(−4, −1) (−3, 0) (−2, 1) (−1, 2) (0, 3) (1, 4) (2, 5) (3, 6)

The number of points that satisfy the equation is infinitely large. You just need a few to draw a decent graph. (Actually, you only need two points to draw a particular line, but I like to graph more for accuracy's sake.) Figure 12-6 shows you the points graphed and then connected to form the line.

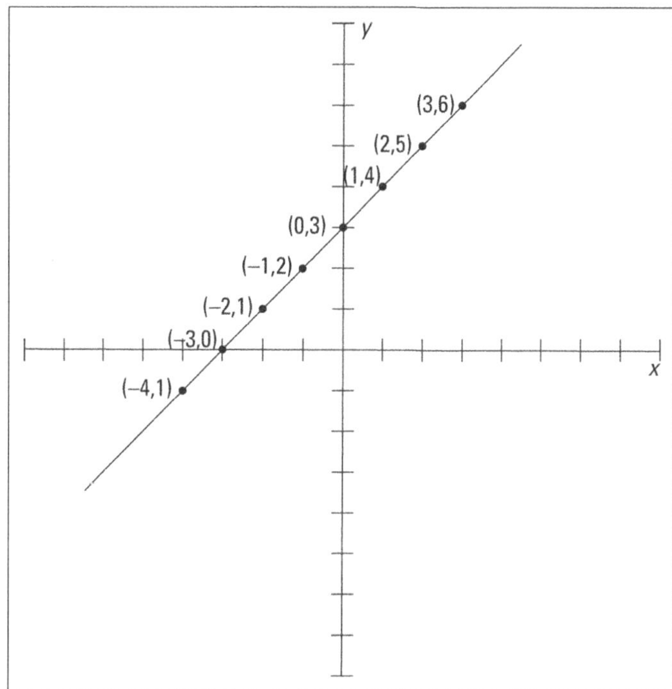

Figure 12-6:
Graphing
$y = x + 3$.

Going around in circles with a circular graph

An example of an equation of a circle is $x^2 + y^2 = 25$. The circle representing this equation goes through an infinite number of points. Here are just some of those points:

(0, 5) (0, −5) (5, 0) (−5, 0) (3, 4) (−3, 4) (4, 3) (4, −3) (−3, 4) (−3, −4) (−4, −3)

I haven't finished all the possible points with integer coordinates, let alone points with fractional coordinates, such as $\left(\frac{25}{13}, \frac{60}{13}\right)$.

When graphing an equation, you don't expect to find all the points. You just want to find enough points to help you sketch in all the others without naming them.

In Figure 12-7, I show you the graph of the circle and some of the named points that make up the graph of the circle.

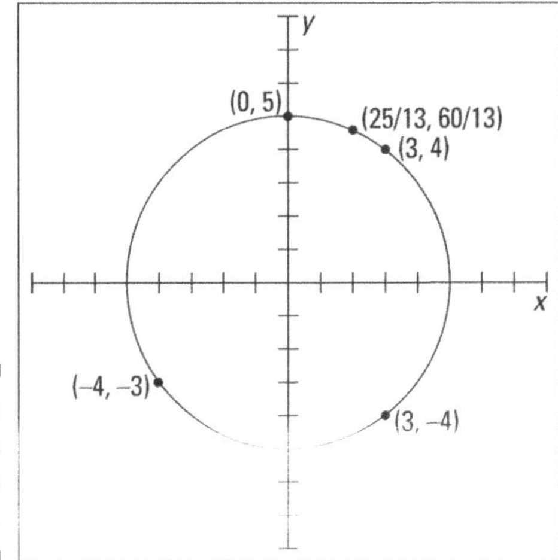

Figure 12-7:
The circle has a radius of 5.

Throwing an object into the air

The equation for the height of an object that's tossed into the air with an initial velocity of v_0 and an initial height of h_0 is $h(t) = -16t^2 + v_0 t + h_0$ where t is the amount of time since the launching of an object. Replacing the t with x and the $h(t)$ with y, I can graph the equation on the coordinate axes.

For example, say a ball is thrown into the air at an initial velocity of 132 metres per second. The person throwing the ball is standing on a building 40 metres tall. So the equation representing the height of the ball is $h(t) = -16t^2 + 132t + 40$ or $y = -16x^2 + 132x + 40$. Graph the equation.

First, compute some of the points by putting in values for x. Starting with 0, and going up by 1, you get the following points:

(0, 40) (1, 156) (2, 240) (3, 292) (4, 312)

(5, 300) (6, 256) (7, 180) (8, 72) (9, −68)

Figure 12-8 shows you the graph and some of the points labelled (see Chapter 14 for more on *parabolas*, which is the shape of the curve shown in Figure 12-8).

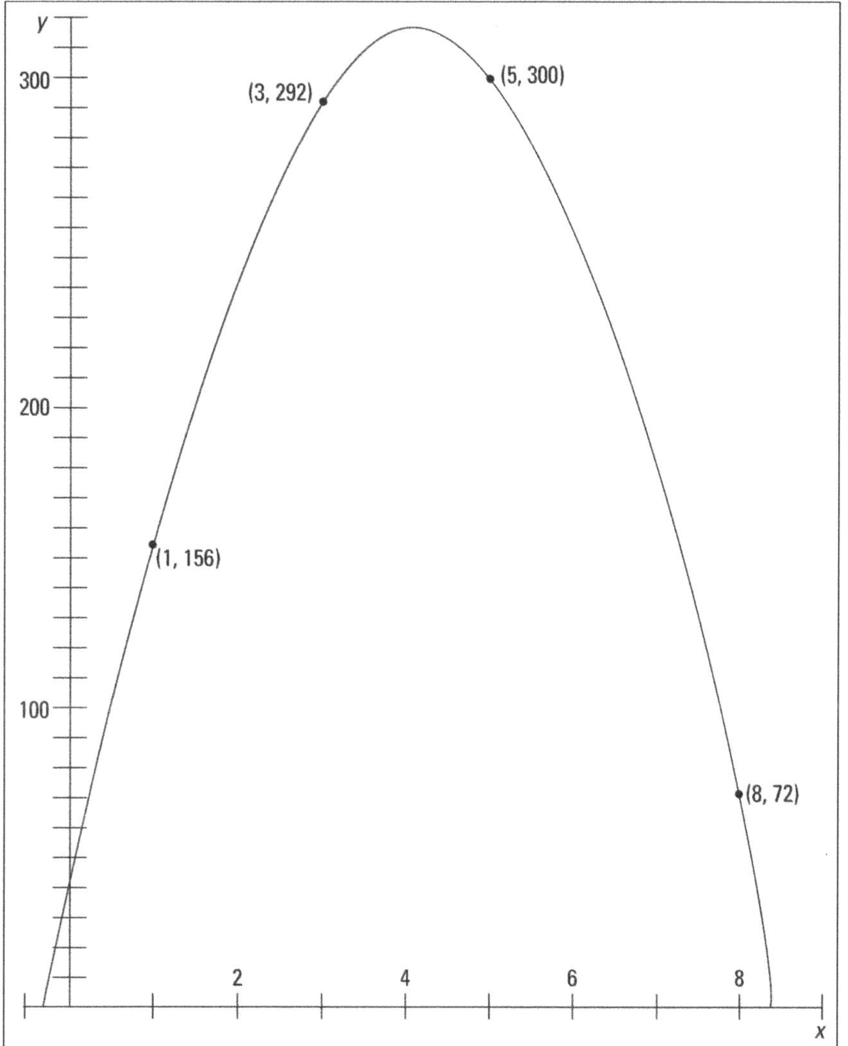

Figure 12-8: He shot an arrow into the air, and where it landed …

Chapter 13

Graphing Lines, Gradients and Circles

In This Chapter

▶ Graphing lines

▶ Sliding down the slippery slope with the gradient formula

▶ Putting perpendicular and parallel lines in the picture

▶ Using the distance formula and finding the halfway mark

▶ Finding the equation for circles

*L*ines are found all around you: 'Get in line!' and 'Toe the line'. You have mental pictures of lines when you hear those commands. But a line is more than a geometric figure and a place to put your feet. Lines are good representations of some activities that go on around you and affect your life. The formulas for determining how much income tax you'll pay in your first part-time job are represented by pieces of lines. The depreciation of goods is often represented by the equation of a line — and its graph shows the decreasing value very vividly.

In this chapter, I present the basics for working with lines and their equations. You find lines determined by two points and then other lines determined by a slope and a point. You see lines that meet and lines that avoid one another forever. The equations of lines are quite straightforward. (Sorry — I couldn't help myself.) I also cover working out the distance and midpoint of a line, and the equation for circles.

Graphing a Line

A straight *line* is the set of all the points on a graph that satisfy a linear equation. When any two points on a line are chosen, the *gradient* (see 'Sighting the Gradient,' later in this chapter) of the segment between those two points is always the same number.

Lines are among the most basic and most useful things to graph in algebra. You can use them to represent how your savings are growing or how a distance from a point changes. They can represent how a piece of machinery depreciates. So, lines are useful, and they're easy to deal with, too. What more could you ask?

Dots or points scattered all over the place with no apparent shape don't usually mean anything. In algebra, it's more common to see points arranged with an equation that gives them something in common. The simplest pattern is a straight line. Line up those points!

The following points are graphed in Figure 13-1:

$$(-2, 12)\quad (-1, 11)\quad (0, 10)\quad (1, 9)\quad (2, 8)\quad (3, 7)\quad (4, 6)\quad (5, 5)$$

$$(6, 4)\quad (7, 3)\quad (8, 2)\quad (9, 1)\quad (10, 0)\quad (11, -1)\quad (12, -2)\quad (13, -3)$$

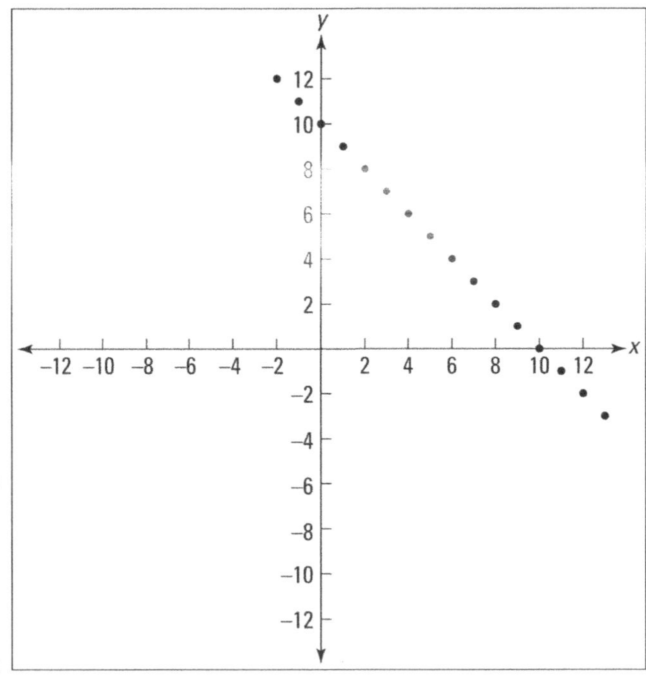

Figure 13-1:
Points lined up like blackbirds, all in a row.

Instead of listing all the millions of points that seem to lie along the same line, you can write an equation that expresses the relation between the points. In the case of Figure 13-1, the relation is $x + y = 10$. The coordinates in each pair add up to 10. But what the graph doesn't show is that not all the points have coordinates that are integers — points such as $\left(1\frac{1}{2}, 8\frac{1}{2}\right)$ that fit the pattern (equation) and lie on the line. By connecting all the points to form a line, you're actually including all the fractional coordinates between the integer coordinate points.

The equation $x + y = 10$ says that any two numbers adding up to 10 give you a point on the graph. This includes fractions, decimals, positives and negatives. What were at first many points or values that worked in the equation are now an infinite number of points.

Figure 13-2 shows you how points look when they're connected to form a line.

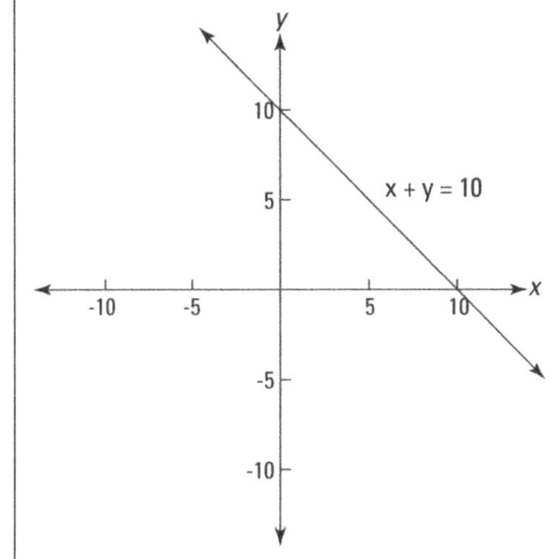

Figure 13-2:
Connect the dots and get a line.

To graph a line, you need only two points. A rule in geometry says that only one line can go through two particular points. Even though only two points are needed to graph a line, it's usually a good idea to graph at least three points to be sure that you graphed the line correctly.

In graphing, three is much better than two. If you get one of two points in the wrong place in a graph, you probably won't notice that the line is wrong. But if you get one of three points in the wrong place, you're more likely to notice that your line isn't straight. Plotting three points is a good check.

Graphing the Equation of a Line

An equation with a graph that's a straight line is said to be *linear*. A linear equation has a standard form of $ax + by = c$, where x and y are variables and a, b and c are real numbers. The equation of a line usually has an x or a y (often both), which refer to all the points (x, y) that make the equation true.

The x and y both have a power of 1. (If the powers were higher or lower than 1, the graph would curve.)

When graphing a line, you can find some pairs of numbers that make the equation true and then connect them. Connect the dots!

What does the equation of a line look like? It looks like any of the following examples. Notice that the first three equations are written in the standard form, and the fourth has you solve for y. The last two have only one variable; this situation happens with horizontal and vertical lines.

$$x + y = 10 \qquad 2x + 3y = 4 \qquad -5x + y = 7$$

$$y = \frac{1}{2}x + 3 \qquad x = 3 \qquad y = -2$$

Whenever you have an equation where y equals a constant number, you have a *horizontal* line going through all those y values. Conversely, if you have an equation where x equals a constant number, all the x values are the same, and you have a *vertical* line. Horizontal lines are all parallel to the x-axis. Their equations look like $y = 3$ or $y = -2$. Vertical lines are all parallel to the y-axis. Their equations all look like $x = 5$ or $x = -11$. Figure 13-3 shows a graph of $y = 4$, using four points: $(-4, 4)$, $(0, 4)$, $(1, 4)$ and $(3, 4)$.

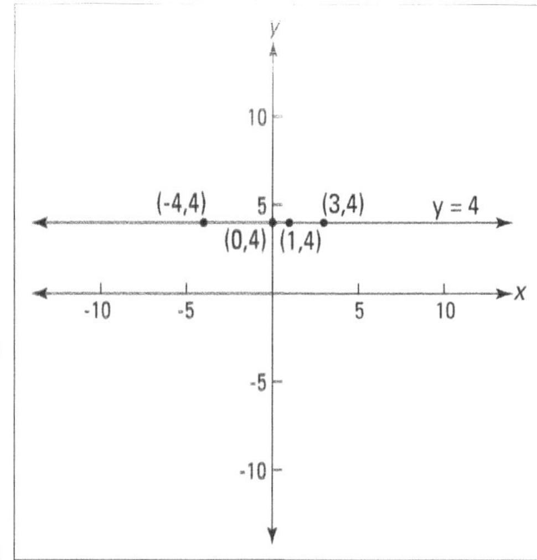

Figure 13-3: Horizontal lines are parallel to the x-axis.

Because a horizontal line is parallel to the x-axis, you might think that the equation of a horizontal line would be $x = a$. And you might figure that the equation for a vertical line would be $y = b$ because a vertical line is parallel to the y-axis. But it's the other way around.

Graphing lines from their equations just takes finding enough points on the line to convince you that you've done the graph correctly.

Try this example. Find a point on the line $x - y = 3$.

1. **Choose a random value for one of the variables, either x or y.**

   To make the arithmetic easy for yourself, pick a large-enough number so that, when you subtract y from that number, you get a positive 3. In $x - y = 3$, you can let $x = 8$, so $8 - y = 3$.

2. **Solve for the value of the other variable.**

   Subtract 8 from each side to get $-y = -5$.

   Multiply each side by -1 to get $y = 5$.

   You can change the looks of the equation without changing the graph of the line by multiplying or dividing each side by the number -1. (For a review of solving linear equations, turn to Chapter 10.)

3. **Write an ordered pair for the coordinates of the point.**

   You chose 8 for x and solved to get $y = 5$, so your first ordered pair is $(8, 5)$.

   You can find more ordered pairs by choosing another number to substitute for either x or y.

For more of a challenge, find points that lie on a line with coefficients on x and y other than 1. The multipliers (2 and 3 in the next example) make this just a little trickier. You may find one or two points fairly easily, but others could be more difficult because of fractions. A good plan in a case like this is to solve for x or y and then plug in numbers.

For example, find points that lie on the line $2x + 3y = 12$.

1. **Solve the equation for one of the variables.**

   Solving for y in the sample problem $2x + 3y = 12$ you get $3y = 12 - 2x$.

   $$y = \frac{12 - 2x}{3}$$

   With multipliers involved, you often get a fraction.

2. **Choose a value for the other variable and solve the equation.**

   Try to pick values so that the result in the numerator is divisible by the 3 in the denominator — giving you an integer.

For example, let $x = 3$. Solving the equation:

$$y = \frac{12 - 2 \cdot 3}{3} = \frac{6}{3} = 2$$

So, the point (3, 2) lies on the line.

Finding the points that lie on the line $x = 4$ may look like a really tough assignment, with only an x showing in the equation. But this actually makes the whole thing much easier. You can write down anything for the y value, as long as x is equal to 4. Some points are: (4, 9), (4, –2), (4, 0), (4, 3.16), (4, –11) and (4, 4).

Notice that the 4 is always the first number. The point (4, 9) is not the same as the point (9, 4). The order counts in ordered pairs.

Graphing these points gives you a nice, vertical line, as Figure 13-4 shows. On the other hand, if all the y-coordinates are the same point, the line is — you guessed it — horizontal.

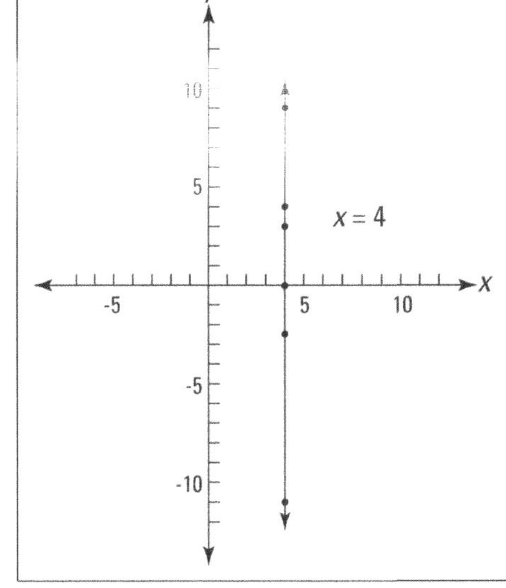

Figure 13-4: When all the x-coordinates are the same, you get a vertical line.

Investigating Intercepts

An *intercept* of a line is a point where the line crosses an axis. Unless a line is vertical or horizontal, it crosses both the x and y axes, so it has two intercepts: An x-intercept and a y-intercept. Horizontal lines have just a y-intercept, and vertical lines have just an x-intercept. The exceptions are when the horizontal line is actually the x-axis or the vertical line is the y-axis.

Intercepts are quick and easy to find and can be a big help when graphing. The reason they're so useful is that one of the coordinates of every intercept is a 0. Zeros in equations cut down on the numbers and the work, and it's nice to take advantage of zeros when you can.

The x-intercept of a line is where the line crosses the x-axis. To find the x-intercept, let the y in the equation equal 0 and solve for x.

For example, find the x-intercept of the line $4x - 7y = 8$.

First, let $y = 0$ in the equation. Then:

$$4x - 0 = 8$$
$$4x = 8$$
$$x = 2$$

The x-intercept of the line is $(2, 0)$: The line goes through the x-axis at that point.

The y-intercept of a line is where the line crosses the y-axis. To find the y-intercept, let the x in the equation equal 0 and solve for y.

For example, find the y-intercept of the line $3x - 7y = 28$. Let $x = 0$ in the equation. Then

$$0 - 7y = 28$$
$$-7y = 28$$
$$y = -4$$

The y-intercept of the line is $(0, -4)$.

As long as you're careful when graphing the x- and y-intercepts and get them on the correct axes, the intercepts are sometimes all you need to graph a line.

Sighting the Gradient

The gradient of a line is a number that describes the steepness and direction of the graph of the line. The gradient is a positive number if the line moves upward from left to right; the gradient is a negative number if the line moves downward from left to right. The steeper the line, the greater the absolute value of the slope (the farther the number is from 0).

Knowing the gradient of a line beforehand helps you graph the line. You can find a point on the line and then use the gradient and that point to graph it. A line with a gradient of 6 goes up steeply. If you know what the line should look like (that is, whether it should go up or down) — information you get from the gradient — you'll have an easier time graphing it correctly.

The value of the gradient is important when the equation of the line is used in modelling situations. For example, in equations representing the cost of so many items, the value of the slope is called the *marginal cost.* In equations representing depreciation, the slope is the *annual depreciation.*

Figure 13-5 shows some lines with their gradients. The lines are all going through the origin just for convenience.

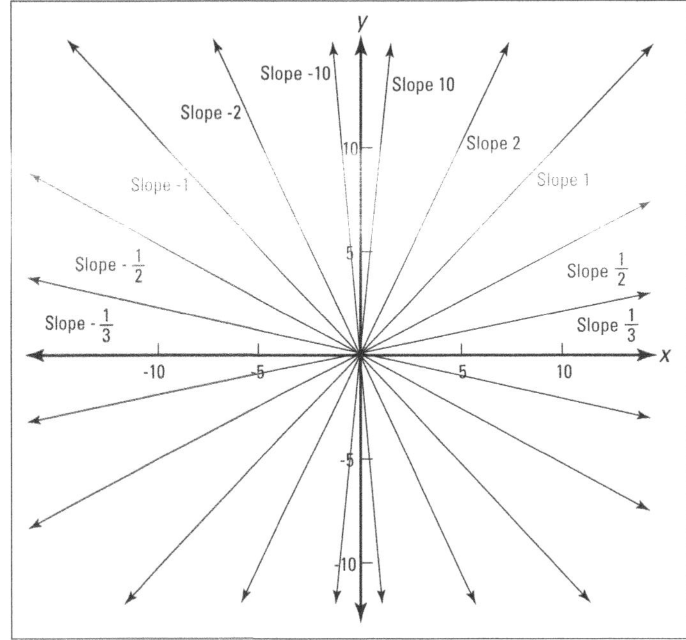

Figure 13-5:
Pick a line
— see its
slope.

What about a horizontal line — one that doesn't go upward or downward? A horizontal line has a 0 slope. A vertical line has no slope; the slope of a vertical line (it's so steep) is undefined. Figure 13-6 shows graphs of lines that have a 0 slope or undefined slope.

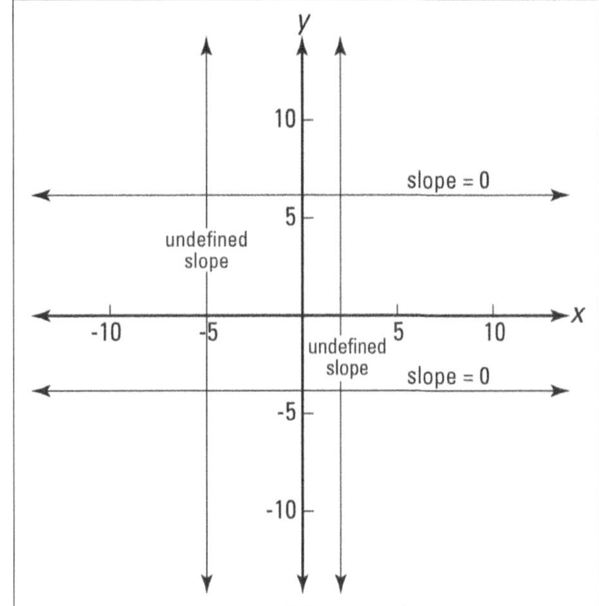

Figure 13-6: Horizontal lines have a 0 slope. Vertical lines have an undefined slope.

One way of referring to the gradient, when it's written as a fraction, is *rise over run*. If the gradient is $\frac{3}{2}$, it means that for every 2 units the line runs along the *x*-axis, it rises 3 units along the *y*-axis. A gradient of $-\frac{1}{8}$ indicates that as the line runs 8 units horizontally, parallel to the *x*-axis, it drops (negative rise) 1 unit vertically.

Formulating gradient

If you know two points on a line, you can compute the number representing the gradient of the line.

The gradient of a line, denoted by the small letter *m*, is found when you know the coordinates of two points on the line, (x_1, y_1) and (x_2, y_2):

$$m = \frac{y_2 - y_1}{x_2 - x_1}$$

Subscripts are used here to identify which is the first point and which is the second point. There's no rule as to which is which; you can name the points any way you want. It's just a good idea to identify them to keep things in order. Reversing the points in the formula gives you the same gradient (when you subtract in the opposite order):

$$m = \frac{y_1 - y_2}{x_1 - x_2}$$

You just can't mix them and do $(x_1 - y_2)$ over $(x_2 - y_1)$.

Now, you can see how to compute gradient with the following examples.

Find the gradient of the line going through (3, 4) and (2, 10).

Let (3, 4) be (x_1, y_1) and (2, 10) be (x_2, y_2). Substitute into the formula:

$$m = \frac{y_2 - y_1}{x_2 - x_2} = \frac{10 - 4}{2 - 3}$$

Simplify:

$$m = \frac{6}{-1} = -6$$

This line is pretty steep as it falls from left to right.

Find the gradient of the line going through (4, 2) and (–6, 2).

Let (4, 2) be (x_1, y_1) and (–6, 2) be (x_2, y_2). Substitute into the formula:

$$m = \frac{y_2 - y_1}{x_2 - x_2} = \frac{2 - 2}{-6 - 4}$$

Simplify:

$$m = \frac{0}{10} = 0$$

These points are both 2 units above the *x*-axis and form a horizontal line. That's why the gradient is 0.

Find the gradient of the line going through (2, 4) and (2, –6).

Let (2, 4) be (x_1, y_1) and (2, –6) be (x_2, y_2). Substitute into the formula:

$$m = \frac{y_2 - y_1}{x_2 - x_2} = \frac{-6 - 4}{2 - 2}$$

Simplify:

$$m = \frac{-10}{0}$$

Oops! You can't divide by 0. There is no such number. The gradient doesn't exist or is undefined. These two points are on a vertical line.

Watch out for these common errors when working with the slope formula:

- ✔ **Be sure that you subtract the *y* values on the top of the division formula.** A common error is to subtract the *x* values on the top.

- ✔ **Be sure to keep the numbers in the same order when you subtract.** Decide which point is first and which point is second. Then take the second *y* minus the first *y* and the second *x* minus the first *x*. Don't do the top subtraction in a different order from the bottom.

Combining gradient and intercept

An equation of a single line can take many forms. Just as you can solve for one variable or another in a formula, you can solve for one of the variables in the equation of a line. This change of format can help you find the points to graph the line or find the gradient of a line.

A common and popular form of the equation of a line is the *gradient-intercept form*. It's given this name because the gradient of the line and the *y*-intercept of the line are obvious on sight. When a line is written $6x + 3y = 5$, you can find points by plugging in numbers for *x* or *y* and solving for the other coordinate. But, by using methods for solving linear equations (refer to Chapter 10), the same equation can be written $y = -2x + \frac{5}{3}$, which tells you that the gradient is –2 and the place the line crosses the *y*-axis (the *y*-intercept) is $\left(0, \frac{5}{3}\right)$.

Where *y* and *x* represent points on the line, *m* is the gradient of the line, and *c* is the *y*-intercept of the line; the gradient-intercept form is $y = mx + c$.

In every case shown next, the equation is written in the gradient-intercept form. The coefficient of *x* is the gradient of the line and the constant is the *y*-intercept:

- ✔ $y = 2x + 3$: The gradient is 2; the *y*-intercept is (0, 3).

- ✔ $y = \frac{1}{3}x - 2$: The gradient is $\frac{1}{3}$; the *y*-intercept is (0, –2).

- ✔ $y = 7$: The gradient is 0; the *y*-intercept is (0, 7). You can read this equation as being $y = 0 \times x + 7$.

Getting to the gradient-intercept form

If the equation of the line isn't already in the gradient-intercept form, solving for y changes the equation to gradient-intercept form.

For example, put the equation $5x - 2y = 10$ in gradient-intercept form:

1. **Get the y term by itself on the left.**

   Subtract $5x$ from each side to get the y term alone: $-2y = -5x + 10$.

2. **Solve for y.**

   Divide each side by -2 and simplify the two terms on the right.

$$\frac{-2y}{-2} = \frac{(-5x+10)}{2}$$

$$y = \frac{-5x}{-2} + \frac{10}{-2}$$

$$y = \frac{5}{2}x - 5$$

The gradient is $\frac{5}{2}$ and the y-intercept is at $(0, -5)$.

Graphing with gradient-intercept

One advantage to having an equation in the gradient-intercept form is that graphing the line can be a fairly quick task, as the following examples show.

Graph $y = \frac{3}{2}x + 1$.

The gradient of this line is $\frac{3}{2}$, and the y-intercept is the point $(0, 1)$. First, graph the y-intercept (see Figure 13-7). Then use the rise-over-run interpretation of gradient to count spaces to another point on the line. To do this, do the run, or bottom, movement first. In this sketch, move 2 units to the right of $(0, 1)$. From there, rise or go up 3 units, which should get you to $(2, 4)$.

It's sort of like going on a treasure hunt: 'Two steps to the east; three steps to the north; now dig in!' Only your 'dig in' is to put a point there and connect that point with the starting point — the intercept. Look at the right side (the b side) of Figure 13-7 to see how it's done.

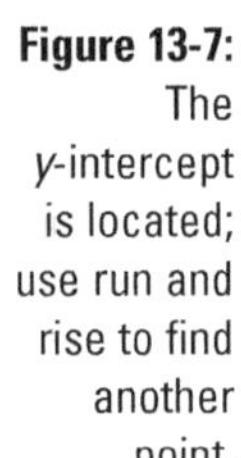

Figure 13-7:
The
y-intercept
is located;
use run and
rise to find
another
point.

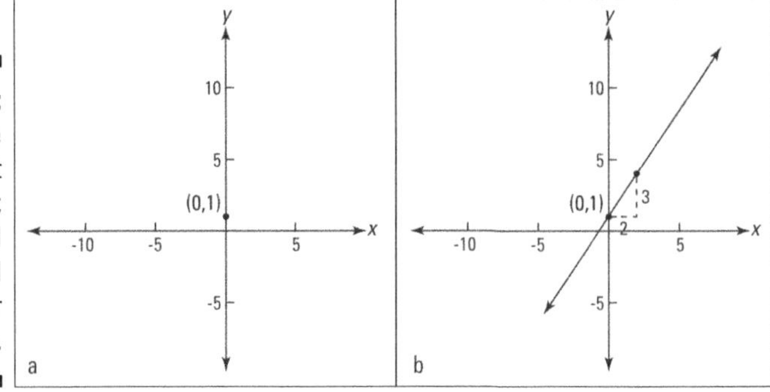

Using a point and the gradient is a quick-and-easy way to sketch a line, so I'll show it to you one more time.

Graph $y = -3x + 2$.

First, graph the *y*-intercept $(0, 2)$. Think of the gradient -3 as being the fraction $-\frac{3}{1}$. This way, you have a run of 1. The rise isn't a rise in this case. The 3 is negative, so it's a fall. Connect the intercept $(0, 2)$ with the point that you find by moving 1 unit to the right and 3 units down, which should be $(1, -1)$. Figure 13-8 shows the line $y = -3x + 2$, which has a gradient of -3.

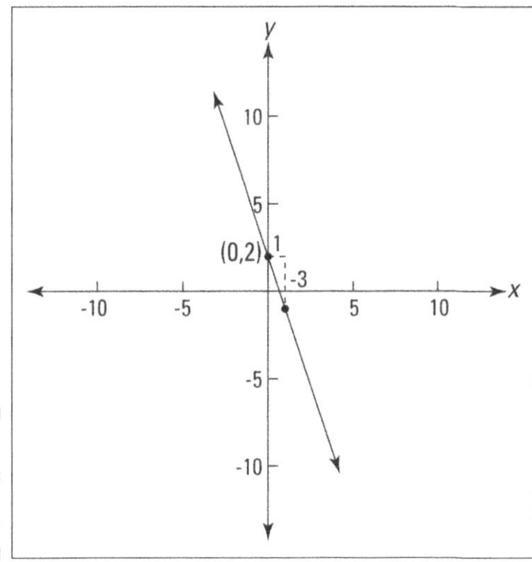

Figure 13-8:
The graph of
$y = -3x + 2$.

Marking Parallel and Perpendicular Lines

The gradient of a line gives you information about a particular characteristic of the line. It tells you if it's steep or flat and if it's rising or falling as you read from left to right. The gradient of a line can also tell you if one line is parallel or perpendicular to another line. Figure 13-9 shows parallel and perpendicular lines.

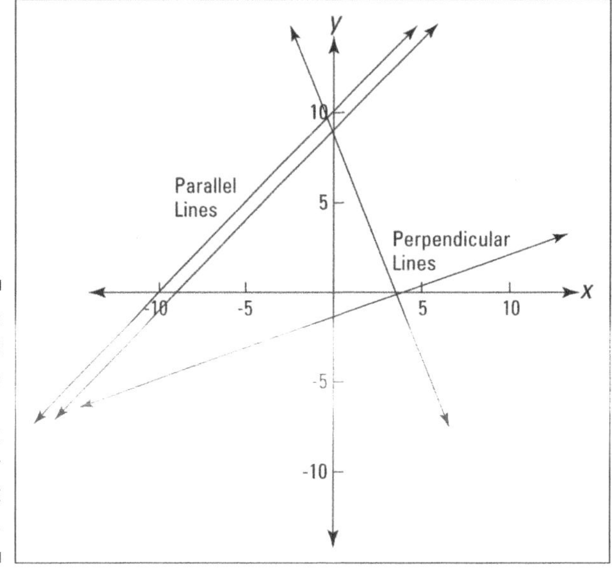

Figure 13-9:
Parallel lines are like railroad tracks; perpendicular lines meet at a right angle.

Parallel lines never touch. They're always the same distance apart and never share a common point. They have the same gradient.

Perpendicular lines form a 90-degree angle (a *right angle*) where they cross. They have gradients that are negative reciprocals of one another. The *x* and *y* axes are perpendicular lines.

Two numbers are reciprocals if their product is the number 1. The numbers $\frac{3}{4}$ and $\frac{4}{3}$ are reciprocals. Two numbers are negative reciprocals if their product is the number −1. The numbers $\frac{3}{4}$ and $-\frac{4}{3}$ are negative reciprocals.

If line y_1 has a gradient of m_1, and if line y_2 has a gradient of m_2, the lines are parallel if $m_1 = m_2$. If line y_1 has a gradient of m_1, and if line y_2 has a gradient of m_2, the lines are perpendicular if $m_1 = -\frac{1}{m_2}$.

The following examples show you how to determine whether lines are parallel or perpendicular by just looking at their gradients:

- The line $y = 3x + 2$ is parallel to the line $y = 3x - 7$ because their gradients are both 3.

- The line $y = -\frac{1}{4}x + 3$ is parallel to the line $y = -\frac{1}{4}x + 1$ because their gradients are both $-\frac{1}{4}$.

- The line $3x + 2y = 8$ is parallel to the line $6x + 4y = 7$ because their gradients are both $-\frac{3}{2}$. Write each line in the gradient-intercept form to see this: $3x + 2y = 8$ can be written $y = -\frac{3}{2}x + 4$ and $6x + 4y = 7$ can be written $y = -\frac{3}{2}x + \frac{7}{4}$.

- The line $y = \frac{3}{2}x + 5$ is perpendicular to the line $y = -\frac{4}{3}x + 6$ because their gradients are negative reciprocals of one another.

- The line $y = -3x + 4$ is perpendicular to the line $y = \frac{1}{3}x - 8$ because their gradients are negative reciprocals of one another.

Intersecting Lines and Simultaneous Equations

If two lines *intersect,* or cross one another, they intersect exactly once and only once. The place they cross is the point of intersection and that common point is the only one both lines share. Careful graphing can sometimes help you to find the point of intersection.

The point (5, 1) is the point of intersection of the two lines $x + y = 6$ and $2x - y = 9$ because the coordinates make each equation true:

- If $x + y = 6$, substituting the values $x = 5$ and $y = 1$ give you $5 + 1 = 6$, which is true.

- If $2x - y = 9$, substituting the values $x = 5$ and $y = 1$ give $2 \times 5 - 1 = 10 - 1 = 9$, which is also true.

This is the only point that works for both the lines.

Graphing for intersections

Careful graphing can give you the intersection of two lines. The only problem is that if your graph is even a little off, you can get the wrong answer. Also, if the answer has a fraction in it, it's difficult to figure out what that fraction is.

For example, find the intersection of the lines $3x - y = 5$ and $x + y = -1$.

Look at the graphs in Figure 13-10. The lines appear to cross at the point $(1, -2)$. Replace the coordinates in the equations to check this out:

- ✔ If $3x - y = 5$, substituting the values gives $3 \times 1 - (-2) = 3 + 2 = 5$, which is true.

- ✔ If $x + y = -1$, substituting the values gives $1 + (-2) = -1$, which is also true.

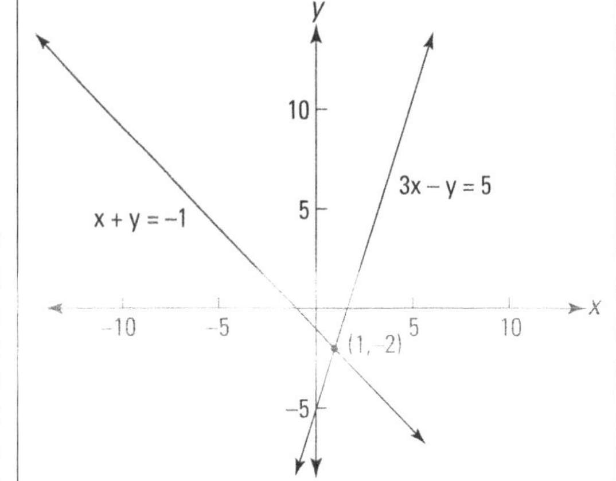

Figure 13-10
The intersection of two lines at a point $(1, -2)$.

Graphing is an inexact way to find the intersection of lines. You have to be super-careful when plotting the points and lines.

Substituting to find intersections

Another way to find the point where two lines intersect is to use a technique called *substitution* — you substitute the y value from one equation for the y value in the other equation and then solve for x. Because you're looking for the place where x and y of each line are the same — that's where they intersect — you can write the equation $y = y$, meaning that the y from the first line is equal to the y from the second line. Replace the ys with what they're equal to in each equation, and solve for the value of x that works.

For example, find the intersection of the lines $3x - y = 5$ and $x + y = -1$. (This is the same problem graphed in the preceding section.) As follows:

1. **Put each equation in the slope-intercept form, which is a way of solving each equation for y.**

   $3x - y = 5$ is written as $y = 3x - 5$, and $x + y = -1$ is written as $y = -x - 1$. The lines are not parallel, and their slopes are different, so there will be a point of intersection.

2. **Set the y points equal and solve.**

   From $y = 3x - 5$ and $y = -x - 1$, you substitute what y is equal to in the first equation with the y in the second equation: $3x - 5 = -x - 1$.

3. **Solve for the value of x.**

   Add x to each side and add 5 to each side:

   $$3x + x - 5 + 5 = -x + x - 1 + 5$$

   $$4x = 4$$

   $$x = 1$$

   Substitute that 1 for x into either equation to find that $y = -2$. The lines intersect at the point $(1, -2)$.

This technique is how the solution can be found without even graphing. If the lines are parallel, it's apparent immediately because their slopes are the same. If that's the case, stop — there's no solution. Also, if the two equations are just two different ways of naming the same line, this will be apparent: The equations will be exactly the same in the gradient-intercept form.

Here's another example. Say you're given the following and asked to determine the value for x and y as the point of intersection: $3x + y = 7$ and $x = -3 - 3y$. You can also solve these types of equations by substituting one into the other. Notice that the first equation has two variables (x and y). You know that you can't solve an equation that has two variables in it, but the second equation tells you that x is equal to an expression that only has y in it.

You can use this to replace the x in the first equation with the expression from the second equation, as follows.

$$3x + y = 7 \text{ and } x = -3 - 3y \text{ means } 3(-3 - 3y) + y = 7$$

When you're replacing a single variable with an expression, pop a bracket around the whole expression. This way you remember to take the whole expression, and that you need to multiply everything in that expression by the coefficient if there is one, like in this example.

Now you only have one variable in the first equation and you solve that quite easily.

$$3(-3-3y)+y=7$$
$$-9-9y+y=7$$
$$-9-8y=7$$
$$-8y=7+9$$
$$-8y=16$$
$$y=\frac{16}{-8}$$
$$y=-2$$

You now find the coordinate value for x to find the intersection point on a graph.

Because you now know the value for y, you can substitute this in to find x. Again, use brackets to keep track of those pesky negative signs.

$$x=(-3-3y), y=-2$$
$$x=(-3-3(-2))$$
$$x=(-3--6)$$
$$x=-3+5$$

Therefore, you can say that the point of intersection is (3, –2).

Eliminating to find intersections

Say you're given the following pair of equations and asked to find the point of intersection: $x+2y=5$ and $-x+4y=1$. Notice that this pair of equations is written quite differently — this is a clue to help you decide to use a different method.

The coefficient of x in the first equation is 1 and in the second it is –1. Of course, when you add the inverse of a number to the actual number, they cancel each other out. So if you add these equations together, you're left with only one variable to solve for. From there, the method is the same as for substitution (refer to preceding section).

$$x + 2y = 5$$
$$-x + 4y = 1$$
$$+$$
$$6y = 6$$
$$y = \frac{6}{6}$$
$$y = 1$$

Again, it doesn't matter which of the original equations you chose to now find x. The answer will be the same. In the following workings, I've used the first equation.

$$x + 2y = 5, y = 1$$
$$x + 2(1) = 5$$
$$x + 2 = 5$$
$$x = 5 - 2$$
$$x = 3$$

So the lines intersect at (3, 1).

Sometimes you need to multiply one, or both of the equations to get them to a form where one of the variables is cancelled out when you add the equations. This may mean you make two new equations to solve for the first variable — just make sure you always go back to the first equation to calculate the second.

For example, say you're given $-3x + 2y = 19$ and $4x + 5y = 13$. You can multiply the first equation by 4 and the second by 3 to end up with $-12x + 8y = 76$ and $12x + 15y = 39$.

Now the coefficients of x are the inverse of each other, so you can add the equations and they can cancel out, allowing you to solve for y.

$$23y = 115$$
$$y = \frac{115}{23}$$
$$y = 5$$

Now you can go back to the first equation and solve for x.

$$-3x + 2y = 19, y = 5$$
$$-3x + 2(5) = 19$$
$$-3x + 10 = 10$$
$$-3x = 19 - 10$$
$$-3x = 9$$
$$x = \frac{9}{-3}$$
$$x = -3$$

The point where these lines intersect is at (3, 1).

Applications of simultaneous equations

Sometimes worded questions are really simultaneous equations in disguise. Once you find the equations and decide which process works best, the method is the same.

For example, say you're told that the difference between two numbers is 10 and their sum is 24. You're then asked to work out what the two numbers are.

Rewrite these as two equations, with the numbers you need to find represented by x and y respectively.

$$x - y = 10$$
$$x + y = 24$$

Now have a look at the coefficients of y. If you use the elimination method from the preceding section and add these equations, you can solve the problem! Too easy!

$$x - y = 10$$
$$x + y = 24$$
$$+$$
$$2x = 34$$
$$x = \frac{34}{2}$$
$$x = 17$$

Once you have solved for x, you can back to the first equation to solve for y.

$$x - y = 10, \; x = 17$$
$$(17) - y = 10$$
$$-y = 10 - 17$$
$$-y = -7$$
$$\frac{-y}{-1} = \frac{-7}{-1}$$
$$y = 7$$

So the two numbers are 17 and 7.

Don't be scared of simultaneous equations! Instead, simply keep the following in mind:

- ✔ Look for clues as to which method will work.

- ✔ Solve for one variable.

- ✔ Substitute the variable you have found into the first equation to find the second variable.

- ✔ Write the answer, the solution, as a coordinate point! Unless it is a worded question where you should always write your answer as a sentence.

Working Out Distance and the Midpoint

If you have two points in the Cartesian plane, as well as working out the gradient between them (refer to the preceding sections), the next two most basic questions you can ask about the points are these:

- ✔ What's the distance between them?
- ✔ What's the location of the point halfway between them?

The distance formula

Ready for your next formula? Here goes (and see Figure 13-11, which illustrates the formula).

The distance between two points (x_1, y_1) and (x_2, y_2) is given by the following formula:

$$\text{Distance} = \sqrt{(x_2 - x_1)^2 + (y_2 - y_1)^2}$$

Don't mix up the slope formula with the distance formula. Both formulas involve the expressions $(x_2 - x_1)$ and $(y_2 - y_1)$. That's because the lengths of the legs of the right-angle triangle in the distance formula are the same as the *rise* and the *run* from the slope formula. To keep the formulas straight, just focus on the fact that slope is a *ratio* of rise over run and that the distance formula *squares* things (like the Pythagorean theorem, because it gives you the length of a hypotenuse — see Chapter 15 for more on this).

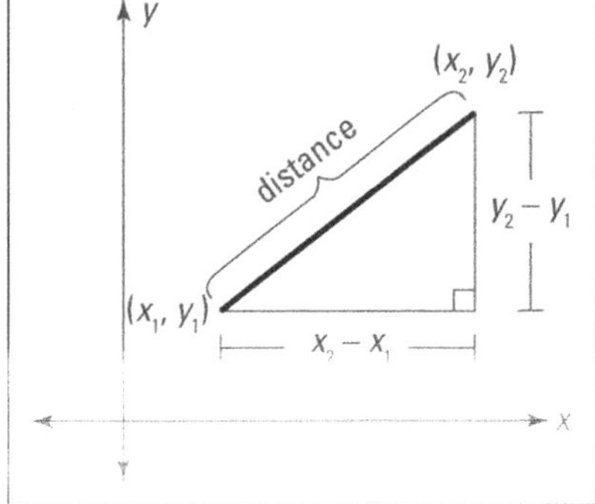

Figure 13-11: The distance formula for working out the distance between two points.

The midpoint formula

The midpoint formula takes the *average* of the *x*-coordinates of the segment's endpoints and the average of the *y*-coordinates of the endpoints.

To find the midpoint of a segment with endpoints at (x_1, y_1) and (x_2, y_2), use the following formula:

$$\text{Midpoint} = \left(\frac{x_1 + x_2}{2}, \frac{y_1 + y_2}{2} \right)$$

Equations for Circles

While you may be used to dealing with graphing lines in the coordinate system, graphing circles may be new for you. No need to worry — you'll soon see that there's nothing to it.

To find the equation of a circle, you need to know its centre point and its radius (see Chapter 15 for more on measurements and circles). You can then use the following formulas:

✔ **For a circle centred at the origin, (0, 0):**

$$x^2 + y^2 = r^2$$

✔ **For a circle centred at any point (h, k):**

$$(x - h)^2 + (y - k)^2 = r^2$$

where (h, k) is the centre of the circle and r is its radius.

Ready for a circle problem? Here you go:

Given: Circle C has its centre at $(4, 6)$ and is tangent to a line at $(1, 2)$

Find: 1. The equation of the circle

2. The circle's x-and y-intercepts

3. The equation of the tangent line

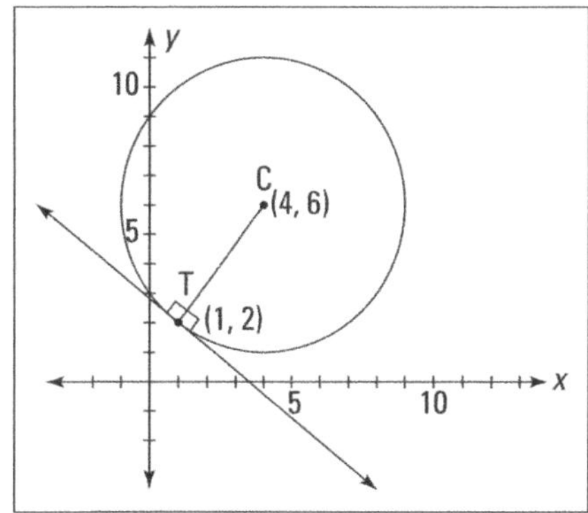

1. Find the equation of the circle.

All you need for the equation of a circle is its centre (you know it) and its radius. The radius of the circle is just the distance from its centre to any point on the circle, but since the point of tangency is given, that's the easiest point to use. This means:

$$\text{Distance}_{C\,to\,T} = \sqrt{(4-1)^2 + (6-2)^2}$$
$$= \sqrt{3^2 + 4^2}$$
$$= 5$$

Now you finish by plugging the centre coordinates and the radius into the general circle equation:

$$(x - h)^2 + (y - k)^2 = r^2$$
$$(x - 4)^2 + (y - 6)^2 = 5^2$$

2. **Find the circle's *x*- and *y*-intercepts.**

   To find the *x*-intercepts for any equation, you just plug in 0 for *y* and solve for *x*:

   $$(x-4)^2 + (y-6)^2 = 5^2$$
   $$(x-4)^2 + (0-6)^2 = 5^2$$
   $$(x-4)^2 + 36 = 25$$
   $$(x-4)^2 = -11$$

   You can't square something and get a negative number, so this equation has no solution; therefore, the circle has no *x*-intercepts. (I realise that you can just look at the figure and see that the circle doesn't intersect the *x*-axis, but I wanted to show you how the maths confirms this.)

   To find the *y*-intercepts, plug in 0 for *x* and solve:

   $$(0-4)^2 + (y-6)^2 = 5^2$$
   $$16 + (y-6)^2 = 25$$
   $$(y-6)^2 = 9$$
   $$y-6 = \pm\sqrt{9}$$
   $$y = \pm 3 + 6$$
   $$y = 3 \text{ or } 9$$

   Thus, the circle's *y*-intercepts are (0, 3) and (0, 9).

3. **Find the equation of the tangent line.**

   For the equation of a line, you need a point (you have it) and the line's slope. Since a tangent line is perpendicular to the radius drawn to the point of tangency, you just compute the slope of the radius, and then the opposite reciprocal of that is the slope of the tangent line.

   $$\text{Slope} = \frac{y_2 - y_1}{x_2 - x_1}$$

   $$\text{Slope}_{\text{Radius } \overline{CT}} = \frac{6-2}{4-1} = \frac{4}{3}$$

   Therefore:

   $$\text{Slope}_{\text{Tangent line}} = -\frac{3}{4}$$

   Now you plug this slope and the coordinates of the point of tangency into the point-slope form for the equation of a line:

   $$y - y_1 = m(x - x_1)$$
   $$y - 2 = -\frac{3}{4}(x - 1)$$

Chapter 14

Getting Familiar with Functions

In This Chapter

▶ Working with parabolas

▶ Getting serious about algebra and functions

▶ Becoming familiar with function families

▶ Understanding how you can solve two equations at the same time

*M*any mathematicians would agree that the most important objects in mathematics aren't numbers; they're functions. Mathematicians use the idea of a *function* to describe operations such as addition and multiplication, transformations of geometric figures, relationships between variables, and many other things. You've probably heard the term *function* in your maths classes at some point.

In this chapter, I explain what you can do with functions, including function families. Firstly, however, I cover parabolas, which often emerge when using functions.

Curling Up with Parabolas

Parabolas are nice, U-shaped curves. They're the graphs of quadratic equations where either an x term is squared or a y term is squared, but not both are squared at the same time. The reflectors in headlights have parabola-like curves running through them. McDonald's golden arches are parabolas. The abundance of manufactured parabolas points to the fact that the properties responsible for creating a parabola often occur naturally. Mathematicians are able to put an equation to this natural phenomenon.

Parabolas have a highest point or a lowest point (or the farthest left or the farthest right) called the *vertex*. The curve is lower on the left and right of a vertex that is the highest point, and it's higher to the left and right of a vertex that is the lowest point.

Trying out the basic parabola

My favourite equation for the parabola is $y = x^2$, the basic parabola. Figure 14-1 shows a graph of this formula. This equation says that the y-coordinate of every point on the parabola is the square of the x-coordinate. Notice that whether x is positive or negative, the y is a square of it and is positive.

The vertex of the parabola in Figure 14-1 is at the origin, (0, 0), and it curves upward.

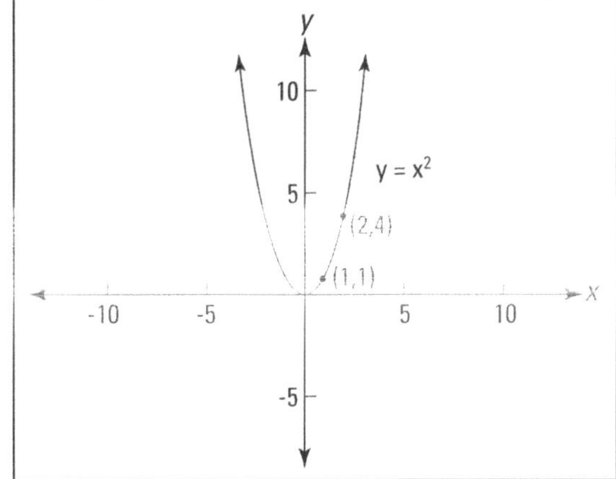

Figure 14-1: The simplest parabola.

You can make this parabola steeper or flatter by multiplying the x^2 by certain numbers. If you multiply the squared term by numbers bigger than 1, it makes the parabola steeper. If you multiply by numbers between 0 and 1 (proper fractions), it makes the parabola flatter (as shown in Figure 14-2). Making it steeper or flatter than the basic parabola helps the parabola fit different applications.

You can make the parabola open downward by multiplying the x^2 by a negative number, and make it steeper or flatter than the basic parabola — in a downward direction.

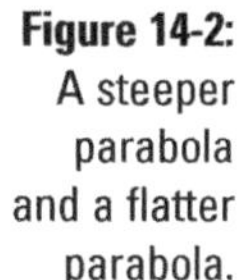
Figure 14-2:
A steeper parabola and a flatter parabola.

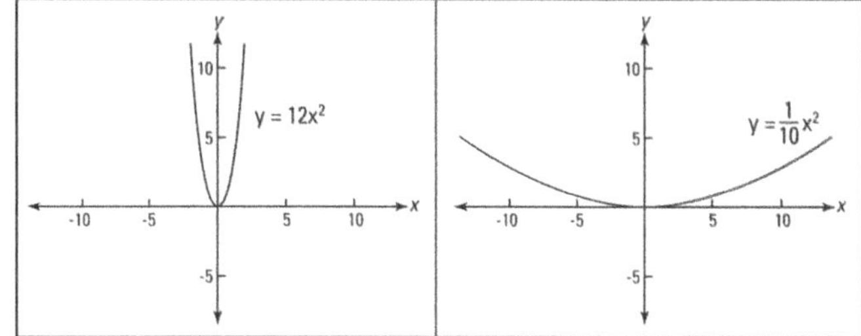

Putting the vertex on an axis

The basic parabola, $y = x^2$, can be slid around — left, right, up, down — placing the vertex somewhere else on an axis and not changing the general shape.

If you change the basic equation by adding a constant number to the x^2 — such as $y = x^2 + 3$, $y = x^2 + 8$, $y = x^2 - 5$, or $y = x^2 - 1$ — then the parabola moves up and down the y-axis. Note that adding a negative number is also part of this rule. These manipulations help make a parabola fit the model of a certain situation.

Note: Not everything starts at 0. Figure 14-3 shows several parabolas, only one of which starts at 0.

If you change the basic parabolic equation by adding a number to the x first and then squaring the expression — such as $y = (x + 3)^2$, $y = (x + 8)^2$, $y = (x - 5)^2$, or $y = (x - 1)^2$ — you move the graph to the left or right of where the basic parabola lies. Using +3 as in the equation $y = (x + 3)^2$ moves the graph to the left, and using –3 as in the equation $y = (x - 3)^2$ moves the graph to the right. It's the opposite of what you might expect, but it works this way consistently (see Figure 14-4).

Sliding and multiplying

You can combine the two operations of changing the steepness of a parabola and moving the vertex. These change the basic parabola to suit your purposes.

The equation used to model a situation may require a steep parabola because the changes happen rapidly, and it may require that the starting point be at 8 metres, not 0 metres. By moving the parabola around and changing the shape, you can get a better fit for the info you want to demonstrate.

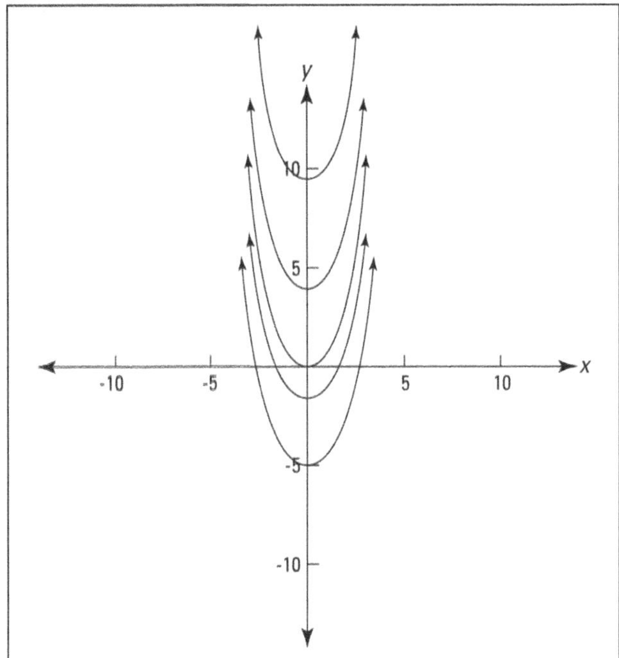

Figure 14-3:
Parabolas
spooning.

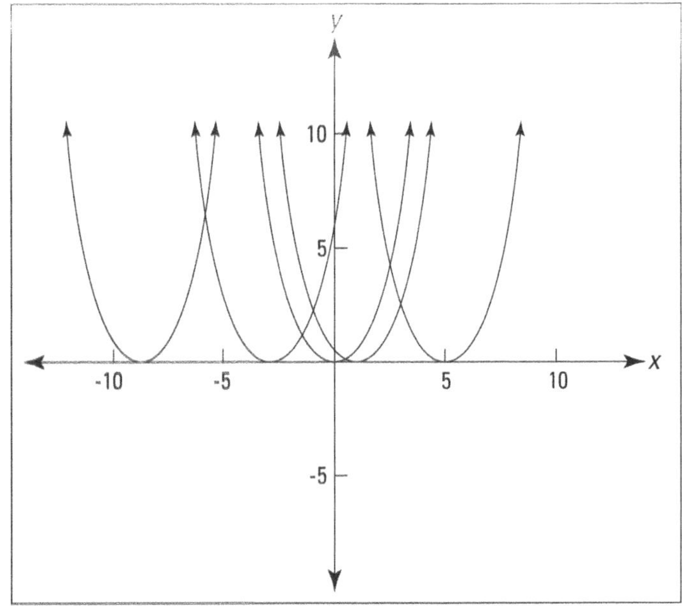

Figure 14-4:
Pretty
parabolas
all in a row.

The following equations and their graphs are shown in Figure 14-5:

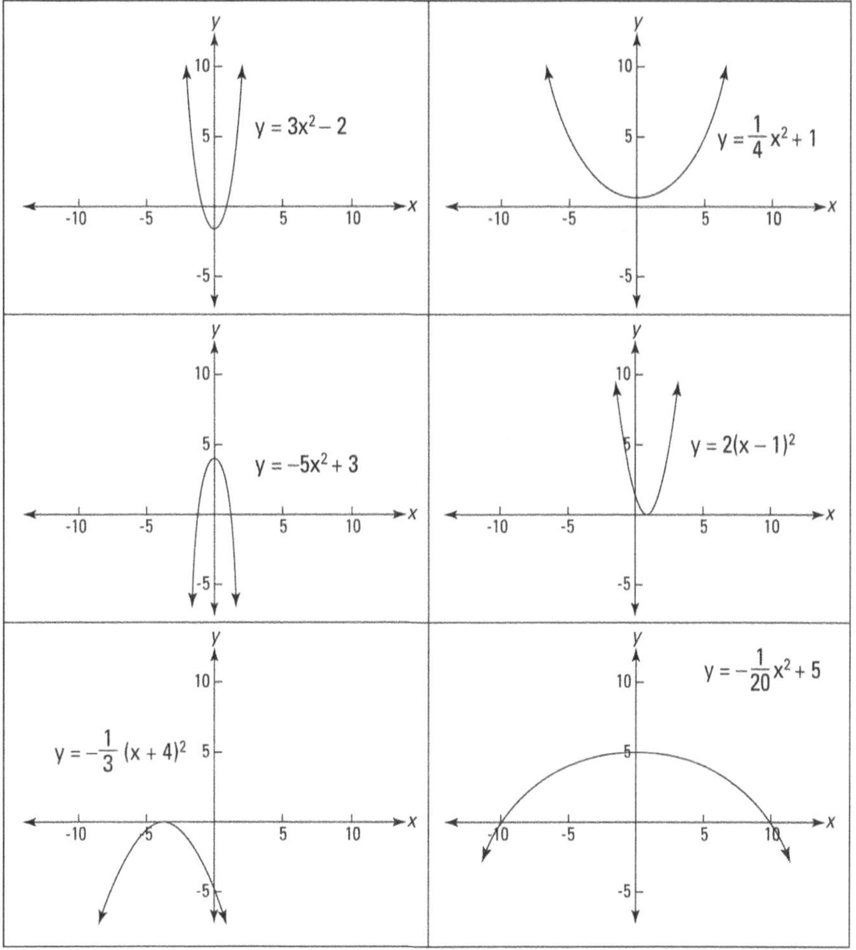

Figure 14-5:
Parabolas
galore.

✔ $y = 3x^2 - 2$: The 3 multiplying the x^2 makes the parabola steeper, and the –2 moves the vertex down to (0, –2).

✔ $y = \frac{1}{4}x^2 + 1$: The $\frac{1}{4}$ multiplying the x^2 makes the parabola flatter, and the +1 moves the vertex up to (0, 1).

✔ $y = -5x^2 + 3$: The –5 multiplying the x^2 makes the parabola steeper and causes it to go downward, and the +3 moves the vertex to (0, 3).

- $y = 2(x - 1)^2$: The 2 multiplier makes the parabola steeper, and subtracting 1 moves the vertex right to (1, 0).

- $y = -\frac{1}{3}(x + 4)^2$: The $-\frac{1}{3}$ makes the parabola flatter and causes it to go downward, and adding 4 moves the vertex left to (–4, 0).

- $y = -\frac{1}{20}x^2 + 5$: The $-\frac{1}{20}$ multiplying the x^2 makes the parabola flatter and causes it to go downward, and the +5 moves the vertex to (0, 5).

Delving into Functions

A *function* is a rule for pairing things up with each other. A function has inputs, it has outputs, and it pairs the inputs with the outputs. This pairing has one important restriction: Each input can be paired with only one output.

An example of something that isn't a function is $y = \pm\sqrt{x}$. In this case, the ± symbol means plus or minus and allows for more than one y-value for the same x-value. This new rule generates these ordered pairs, for example: (4, 2) and (4, –2). The same input (4) is paired with two different outputs (2 and –2), which means that $y = \pm\sqrt{x}$ isn't a function.

On the other hand, the function $y = x^2$ pairs numbers with each other. The x-values are the inputs and the y-values are the outputs. Often, people write the pairs using this notation: (x, y), which means that (1, 1), (2, 4), and (–3, 9) are all pairs generated by the rule $y = x^2$. As required, this function has only one output for each input. There is no requirement the other way around, though. It's okay that (2, 4) and (–2, 4) are both pairs generated by this rule. In this case, one output is paired with two different inputs, which can happen (and it often does) with a function.

Students learn to identify functions and non-functions from their graphs. If two points are on a graph so that one point is directly above the other, it means that the same x-value is paired with two different y-values, so the graph doesn't describe a function. Sometimes, it's referred to as the *vertical line test*. The graph on the left in Figure 14-6 is of $y = x^2$ and the graph on the right is of $y = \pm\sqrt{x}$.

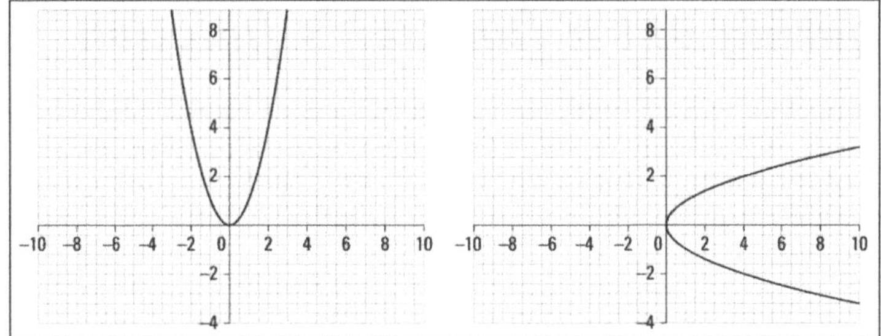

Figure 14-6:
On the left, a function. On the right, not a function.

Understanding the practical side of functions

Knowing the definition of a function is a good thing. In mathematics, definitions help you to sort the world out in a matter-of-fact way. Something either is a function or it isn't a function because it either fits the definition or it doesn't.

Most people, however, don't go through life — or even through maths class — constantly thinking about definitions. Getting better at referring to and using definitions is good mathematical practice, but most people refer first to a set of mental images when they think about categorising things. People tend to ask whether something looks like a function or not before they refer to the definition. Because you may spend so much time with functions such as the one on the left in Figure 14-6, you can miss the strange cases of functions such as those in Figure 14-7.

Each of the graphs in Figure 14-7 represents a function because it fits the definition of pairing *x*-values with *y*-values in such a way that each *x*-value gets only one *y*-value. However, none of the graphs in Figure 14-7 looks like the functions that you likely encounter as you go about your average day in a traditional maths class. But you may encounter (and even produce) these kinds of examples in the service of better understanding functions.

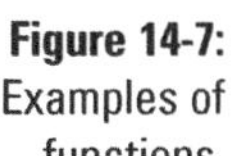

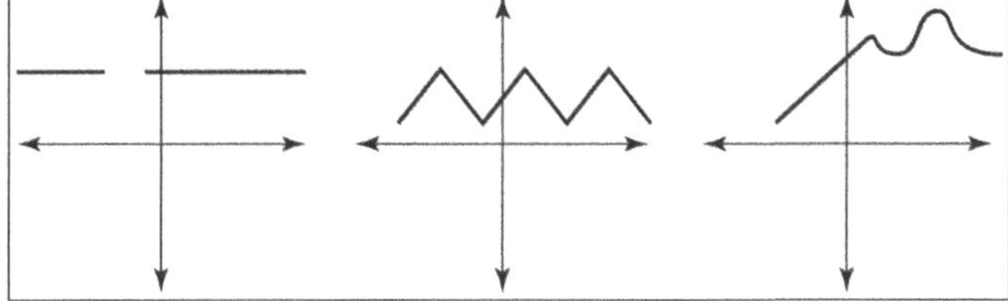

Figure 14-7:
Examples of functions.

Figuring out a function's function

An important way to think about functions is as relationships between variables. If you think of a function as a relationship, you can keep an eye out for useful features. These features include whether (and where) a function is increasing or decreasing, whether a function is linear or not, and so on.

You can read a graph from left to right. If the graph rises as you read from left to right, it means that the y-values are *increasing*. If the graph falls as you read from left to right, the y-values are *decreasing*. If the graph is horizontal, the y-values are *constant*. A function can be increasing for all x-values (such as the line $y = 2x + 2$) or it can be increasing in some places and decreasing in other places (such as $y = x^2 + 2$). See Figure 14-8 as an example.

Figure 14-8: Graph (a) is increasing for all x-values; graph (b) isn't.

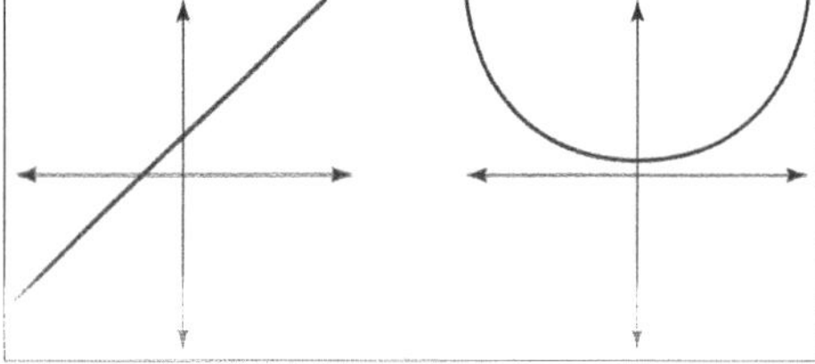

At this stage in your maths classes, you may simply be asked to describe the change you see in graphs of functions.

Studying Function Families

A function is an important mathematical object. At an abstract level, a *function* is a matching of things (usually numbers, but it doesn't have to be). The function $y = x^2$ matches the number 2 to the number 4, for example — when $x = 2$, $y = 4$. As mentioned earlier in this chapter, in order to be a proper *function*, this matching has to be done so that each x-value is matched to only one y-value, although two x-values may both be matched to the same y-value. All of this is terribly abstract, but let me give you the takeaway message: One way to think about functions is as a *matching* (or *correspondence*) of individual numbers.

Another way to think about functions is as *relationships* between variables. As x increases, y increases, for example. Or doubling x also doubles y is another example of a relationship between variables. Every function written as y = some expression or other describes a relationship.

These relationships fall into a number of important and useful categories, and these categories and their characteristics are important objects of study. A few of these categories are

- Linear functions

- Exponential functions

- Quadratic functions and other polynomials

- Rational functions

Each can be considered a *function family*. When you develop familiarity with these families and their properties, you can use various types of functions to describe some commonly appearing applied mathematical situations.

Certain kinds of mathematicians don't like the *relationship* idea of function because, formally, a function doesn't need to have any kind of relationship between the two variables. In fact, some functions have no predictable relationship at all. The function that matches the roll of a six-sided die with its outcome is an example — the first roll may give a 3, a second roll may give a 2, and so on. You don't have a way to know what the next roll will be — so there is no relationship — but the matching is done in a way that fits the definition of function.

In addition to this collection of basic relationships, you can also consider the reverse of these relationships. If y quadruples when x doubles, what happens to x when y doubles? That sort of question involves the idea of *inverse*. Studying inverses introduces whole new families of functions, such as logarithmic functions and roots.

Chapter 15

Pythagoras, Trigonometry and Measurement

In This Chapter

▶ Dealing with units of length

▶ Using the Pythagorean theorem and getting around the perimeter

▶ Working out the area

▶ Determining the volume

▶ Getting fancy with ratios

*Y*ou can't get away from it: Square metres of carpeting, kilometres per litre for the car, capacity of the new freezer. You measure, and you use the appropriate formula to give you the answer you're hoping for.

A formula is a relationship that's proven to be true, no matter what. One of the first formulas that you learned is that the area of a rectangular area is based on how long and how wide.

In this chapter, I reacquaint you with area, perimeter and volume. You also see how to deal with those awkward, irregularly shaped objects. It isn't all that important that you memorise the formulas — the main emphasis is on how to use the formula and where to find it when you need it.

Measuring Up

Some universal concerns — some start at an early age — are those dealing with measurements. How far is it? How big are you? How much room do you need, anyway? How much more wrapping paper are you going to need? These questions all have to do with measurements and, usually, formulas.

Finding out how long: Units of length

Before measurements were standardised, they varied according to who was doing the measuring: A yard was the distance from the tip of the nose to the end of an outstretched arm; a foot, well, you can probably guess where that came from; and an inch was often the length of the second bone in the index finger.

The units of measure for length most commonly used in Australia are centimetres, metres and kilometres. A hundred centimetres make a metre, as the name implies, and these units of measurement are part of the metric system — the same system that uses kilograms and grams for mass.

The idea of the metric system is that converting between units is simply a matter of multiplying or dividing by 10, 100 or 1,000.

You can change the basic length equivalencies into formulas as follows:

- **Metres to centimetres:**

  Number of centimetres = number of metres × 100

- **Centimetres to metres:**

  Number of metres = number of centimetres ÷ 100

- **Kilometres to metres:**

  Number of metres = number of kilometres × 1,000

Putting the Pythagorean theorem to work

Another great formula to use when working with lengths and distances is the Pythagorean theorem. The Pythagorean theorem is a formula that shows the special relationship between the three sides of a right-angled triangle.

A *right-angled triangle* (as opposed to a wrong triangle?) is one with a 90-degree angle.

Pythagoras noticed that if a triangle really was a right-angled triangle, the square of the length of the *hypotenuse* (the longest side) is always equal to the sum of the squares of the two shorter sides:

$$(\text{length of hypotenuse})^2 = (\text{length of a shorter side})^2 + (\text{length of remaining side})^2$$

For example, a triangle with sides measuring 3 centimetres, 4 centimetres and 5 centimetres is a right-angle triangle. The longest side is the one that measures 5 centimetres; the square of 5 is 25. The two shorter sides are 3 and 4 centimetres; $3^2 = 9$ and $4^2 = 16$. The sum of 9 and 16 is 25 — the square of the longest side.

This property works only for right-angle triangles, and if the relationship between the sides works, the triangle has to be a right-angle triangle. Figure 15-1 shows you a general right-angled triangle, as well as my favourite 3-4-5 right-angled triangle.

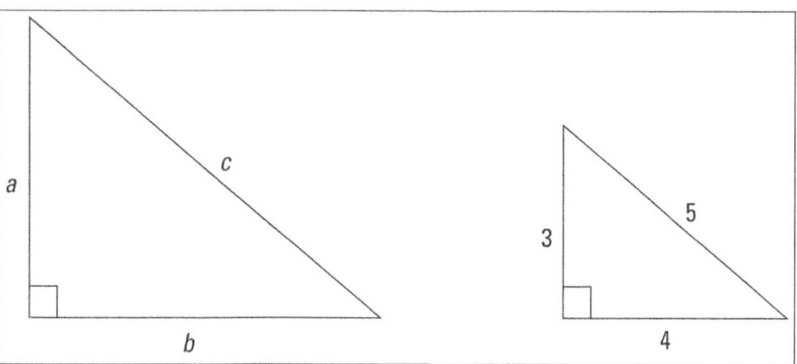

Figure 15-1:
Triangulating the 'right' way.

According to the Pythagorean theorem, if a, b and c are the lengths of the sides of a right-angled triangle, and c is the longest side (the hypotenuse), then $a^2 + b^2 = c^2$.

The following examples show how you can use the Pythagorean theorem.

Show that a triangle with sides that measure 5, 12 and 13 is a right-angled triangle.

First, find the square of the measure of each side:

$$5^2 = 25 \qquad 12^2 = 144 \qquad 13^2 = 169$$

Add the two smaller squares together. That sum is the same as the largest square:

$$25 + 144 = 169$$

Say a carpenter wants to determine whether a garage doorway has square corners or if it's really leaning to one side. She measures 30 centimetres from one corner along the bottom of the doorway and makes a mark. She measures 40 centimetres up along the door frame from the same corner and makes a mark on the side. She then takes a tape measure and measures the distance between the marks; it comes out to be 49 centimetres.

Puzzling Pythagoras

Pythagoras was born somewhere around 570 BC. He is best known for his Pythagorean theorem, but he's also responsible for discovering an important musical property: The notes sounded by a vibrating string depend on the length of the string.

Pythagoras was a great thinker, but he also exhibited some rather bizarre behaviour. He founded a school where about 300 young aristocrats studied mathematics, politics, philosophy, religion, music and astronomy. They formed a very tight fraternity or secret society where the members had their diets and actions regulated. They weren't allowed to eat beans or drink wine or pick up anything that had fallen or stir a fire with an iron. They had to face in a certain direction when they urinated. These strange beliefs supposedly caused Pythagoras's death. When he was being chased from his burning home by some persecutors, he was supposed to have stopped at the edge of a bean field and, rather than trample the beans, allowed his chasers to catch and kill him.

Find the squares of the measures:

$$30^2 = 900 \qquad 40^2 = 1{,}600 \qquad 49^2 = 2{,}401$$

Then, $900 + 1{,}600 = 2{,}500 \neq 2{,}401$. The two smaller squares don't add up to the larger square, so the corner isn't square.

If you're only given measurements for one of the lengths of a right-angle triangle and one of the angles, you can still calculate the other lengths in the triangle. You just need the trigonometric ratios — covered in 'Triggering Trigonometric Ratios', later in the chapter.

Working around the perimeter

How long is the running track around the field? What's the distance around the room? How many metres of fencing does your mum need to go around the pool? The *perimeter* is the distance around the outside of a given figure — the total length of the periphery that borders a region.

In general, the perimeter of a figure is the sum of the lengths of the sides. If you have a triangle, measure each of the three sides and add them up. If you have a four-sided figure, add up the four lengths, and so on. Perimeter formulas are used to simplify this process when you recognise that the figure is something special. You can use quick, easy formulas to do the computations. I give you many of these formulas in this section.

Triangulating triangles

The perimeter of a triangle is equal to the sum of the measures of the three sides — sides s_1, s_2 and s_3: $P = s_1 + s_2 + s_3$.

The formula for the perimeter of a triangle shows the variable s and the subscripts 1, 2 and 3. The s stands for side. Rather than use an a, b and c for the lengths, it's customary to use a single variable (like s, in this case) and a number of subscripts when nothing is special about the sides or how their lengths relate to one another. Subscripts are also used for a figure with more than 26 sides because you can only use a to z to name the sides, but with subscripts you can go on as long as you like — the figure could have a thousand sides. Heaven forbid!

The following examples show you how to find the perimeter of some triangles.

Find the perimeter of the triangle with sides 5 metres, 11 metres and 13 metres.

$$P = 5 + 11 + 13 = 29 \text{ metres}$$

Find the amount of fencing you'll need for a triangular area if the two sides that form a right-angled triangle are 7 metres and 24 metres, and you can't measure the longest side, the hypotenuse, because it's too muddy right now.

Because you have a right-angled triangle, the sum of the squares of 7 and 24 is equal to the square of the longest side:

$$7^2 + 24^2 = 49 + 576 = 625$$

Because 625 is the square of 25, the sides of the area are 7, 24, and 25 metres. Then $P = 7 + 24 + 25 = 56$ metres of fencing needed.

Squaring up to squares and rectangles

A square is wonderful to work with because you have only one measure to worry about — the length of one side is the same as all the others. A rectangle is a special four-sided figure, too. Figure 15-2 shows a rectangle with square (90-degree) corners, where the opposite sides are the same length.

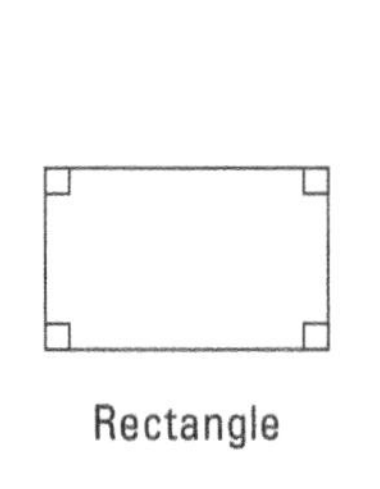

Figure 15-2: A shape for rooms, posters and corrals.

To find the perimeter of a square or rectangle, use the following formulas:

- ✔ The perimeter of a *square* is four times the length of a side: $P = 4s$ (which is easier than adding $s_1 + s_2 + s_3 + s_4$).

- ✔ The perimeter of a rectangle is twice the length plus twice the width. Or you can add the length and width together and then multiply that sum by two. These formulas are easier than adding up the four sides: $P = 2l + 2w = 2(l + w)$ or $P = s_1 + s_2 + s_3 + s_4$.

The following examples illustrate using the formulas for perimeter.

An environmental group is going to search a square kilometre of bushland to check for toxins in beetles. What is the perimeter of that square kilometre in metres?

You know that 1 kilometre is 1,000 metres. So the perimeter is $4 \times 1,000 = 4,000$ metres. So, if they want to rope off the area, they need plenty of rope!

Here's another example. Your nan's new garden is a rectangle measuring 35 metres long by 15 metres wide. You've offered to work out how much fencing she needs to enclose it.

What's the perimeter? Add the 35 and 15 together and double it: $2(35 + 15) = 2(50) = 100$ metres of fencing. Of course, this doesn't include a gate — you should probably consider that, too, unless your nan likes jumping hurdles.

Promoting polygons

A *polygon* is a dead parrot. (Sorry — maths humour tends to have an evil bent to it.) Seriously, a *polygon* is a many-sided figure with the endpoints of each side meeting at the endpoints of the adjacent sides. The sides are all line segments.

In general, the perimeter of a polygon is simply the sum of the measures of the sides. When you have a *regular polygon* (a polygon in which all the sides and all the interior angles are the same), the perimeter is found by multiplying the number of sides, *n*, by the length of any one of those sides, *s*: $P = ns$.

For example, a standard highway stop sign has eight sides that each measure about 31 centimetres. If you want to put a reflective strip all around the outer edges of a stop sign, how many inches is that?

Multiply the length of one side times 8: $P = ns = 8(31) = 248$ centimetres.

Recycling circles

A circle has a perimeter, but this has a special name: *Circumference*. Think about the word: If the *circumstances* (conditions around you) are positive, you can *circumnavigate* (sail around), *circumvent* (go around and avoid), and *circumscribe* (draw around). To find the circumference of a circle, all you need is the measure of the radius or the diameter. The radius is the distance from the centre of the circle to any point on the circle. If you double the radius, you get the measure of the *diameter,* the distance from one side of the circle to the other through the centre point. Figure 15-3 shows a circle with the radius marked. The diameter line would continue through the centre point to the other side of the circle.

Figure 15-3: In a circle, all points are equidistant from the centre.

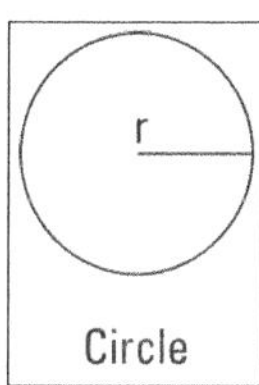

The formula for *circumference* (distance around the outside of a circle) is $C = 2\pi r$ or $C = \pi d$, where *r* is the radius, *d* is the diameter, and π is a constant and can be approximated by 3.14 or $\frac{22}{7}$.

The symbol for the relationship between the circumference and diameter of a circle is the Greek letter π. The value of π is the same — no matter what size circle you have. The decimal value of π is approximately 3.14.

Say you want to construct a circular garden but you're a member of the waste-not-want-not club. The fencing you want comes in bundles of 25 metres, 50 metres, 100 metres, 150 metres, 200 metres and so on, so you're going to construct your garden such that it uses every bit of the fencing around the circumference. How can you easily determine the diameter of each garden with respect to the different fencing amounts?

You should rewrite the formula so you can easily determine how wide your circular garden will be if you buy a certain size bundle of fencing to put around it and use all the fencing in the bundle.

Solving for d in the formula $C = \pi d$, divide each side by π:

$$\frac{C}{\pi} = \frac{\cancel{\pi} d}{\cancel{\pi}}$$

$$\frac{C}{\pi} = d$$

The diameter is equal to the circumference divided by π.

$$d = \frac{C}{\pi}$$

If the bundle has 50 metres of fencing:

$$d = \frac{50}{3.14} \approx 15.92 \text{ metres across}$$

If the bundle has 100 metres of fencing:

$$d = \frac{100}{3.14} \approx 31.85 \text{ metres across}$$

If the bundle has 200 metres of fencing:

$$d = \frac{200}{3.14} \approx 63.69 \text{ metres across}$$

If you know the dimensions of the lot where you're putting your garden, you can determine which garden will fit.

Spreading Out: Area Formulas

Area is a measure of how many two-dimensional units (squares) a particular object or surface covers — how much flat space it occupies. Usually, area is given in square millimetres, square centimetres, square metres or square kilometres and so on.

Picture a floor covered with square tiles. If each tile is 1 metre by 1 metre, counting the number of tiles tells you how large the floor is in square metres. In the real world, most floors aren't covered with tiles that are a convenient 1-metre square. And most tile floors have partial tiles on the edges and in the corners and around things that are strange shapes. So area formulas, such as the ones in this section, help you do the figuring to determine the area.

In the previous section, the perimeter formulas deal with linear measure. Linear measure is just one dimension. It's from one place to another — there's no breadth to it. You measure it with a ruler or tape measure in one direction. Square measurements are used to measure area. Area takes two measures — one along a side and a second perpendicular (90 degrees) to that side.

Laying out rectangles and squares

Rectangles and squares have basically the same area formulas because they both have square corners and the equal lengths on opposite sides. The general procedure here is just to multiply the measure of the length times the measure of the width. The product of two sides that are next to one another is the area.

Finding the area of a rectangle or square

Most rooms in homes, schools and offices are rectangular in shape. Desks and tables and rugs are usually rectangular, also. This makes it easy to fit furniture and other objects in the room.

The area of a rectangle is its length times its width, and the area of a square is the square of the measure of any side:

Rectangle: $A = l \times w$

Square: $A = s^2$

Say a garden 35 metres long by 15 metres wide needs some fertiliser. If a bag of fertiliser covers 6 square metres, how much fertiliser do you need?

First determine how many square metres the garden is.

area of garden $= l \times w = 35 \times 15 = 525$ square metres

Divide the 525 square metres by 6 square metres:

$525 \div 6 = 87.5$ square metres

You can buy 88 bags and have some left over, or buy 87 bags and skimp a little in some places.

Tuning in triangles

Finding the area of a triangle can be a bit of a challenge. Basically, a triangle's area is half that of an imaginary rectangle that the triangle fits into. However, it isn't always easy or necessary to find the length and width of this hypothetical rectangle — you just need a measurement or two from the triangle.

The traditional formula for finding the area of a triangle involves the length of the base, or bottom, and the height, the perpendicular distance from the base up to the vertex (the intersection of the other two sides). Finding the area of a triangle is easy if you can use a ruler to find the height, but that isn't always practical. So, you have another option — Heron's formula, covered later in this section.

Going the traditional route

The area of a triangle is equal to half the product of the measure of the base of the triangle, b, times the height of the triangle, h: $A = \frac{1}{2}bh$.

The base is the length of the bottom that the height is drawn down to. The height is the length from the top angle down perpendicular to the base. The height forms a right angle (90 degrees) with the base. Figure 15-4 shows you a triangle with a height drawn.

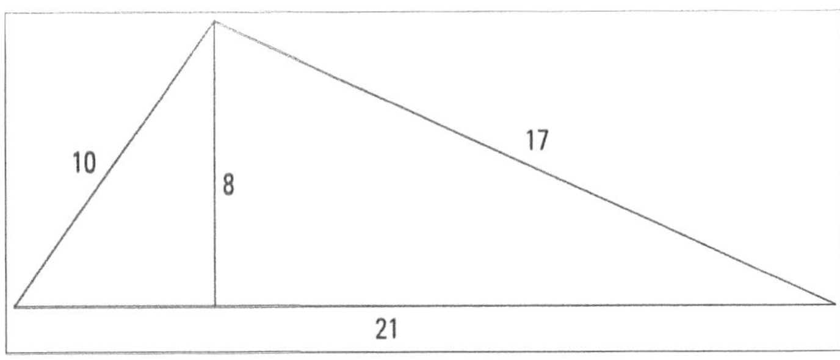

Figure 15-4: Triangles come in all shapes and sizes.

You use this traditional rule for area when it's possible to make these measurements — when you can draw the height perpendicular to the base and measure both of them. The following example shows you how to use the best-known rule first, and a later example finds the same area using Heron's formula.

Find the area of a triangle 21 metres long with a height of 8 metres. Refer to Figure 15-4 for a sketch of such a triangle.

$$A = \frac{1}{2}(21)(8) = \frac{1}{2}(168) = 84 \text{ square metres}$$

Soaring with Heron's formula

According to Heron's formula, the area of any triangle is equal to the square root of the product of four values:

- The semi-perimeter (half the perimeter)
- The semi-perimeter minus the length of the first side
- The semi-perimeter minus the second side
- The semi-perimeter minus the third side

Let s represent the semi-perimeter and a, b and c represent the measures of the sides:

$$A\sqrt{s(s-a)(s-b)(s-c)}$$

When you're trying to find the area of a huge triangle — say a big park — or if you can't measure any angles to draw a line perpendicular to one of the sides for the height, you can find the area simply by measuring the three sides and using Heron's formula.

For example, find the area of a triangle with sides of 10 centimetres, 17 centimetres and 21 centimetres (refer to Figure 15-5).

Let $a = 10$, $b = 17$ and $c = 21$. The perimeter is $P = 10 + 17 + 21 = 48$ centimetres, so the semi-perimeter $s = 24$ centimetres. Using Heron's formula to find the area,

$$A = \sqrt{s(s-a)(s-b)(s-c)} = \sqrt{24(24-10)(24-17)(24-21)}$$
$$= \sqrt{24(14)(7)(3)} = \sqrt{7056} = 84$$

The area is 84 square centimetres. Does that number sound familiar? It should. The 84 part is the same answer as in the previous example where I used the more well-known formula.

I have to admit that I purposely used measurements that would give a nice, whole-number answer. These nice answers are more the exception than the rule. Having a radical in a formula can cause all sorts of complications. The next example shows you what I mean.

Find the area of a triangle with sides 2, 3 and 4 metres. If the sides are 2, 3 and 4, then $a = 2$, $b = 3$, $c = 4$ and $s = \frac{1}{2}(2+3+4) = 4.5$. So, using Heron's formula to find the area,

$$A = \sqrt{s(s-a)(s-b)(s-c)} = \sqrt{4.5(4.5-2)(4.5-3)(4.5-4)} =$$
$$= \sqrt{4.5(2.5)(1.5)(.5)} = \sqrt{8.4375} \approx 2.905 \text{ square metres}$$

Going around in circles

The area of a circle is tied to both the radius of the circle and the value of π.

The formula for the area of a circle is π (about 3.14) times the radius squared: $A = \pi r^2$.

For example, find the area of a circular disc that is 50 metres across. First, you need to find the radius. If the circle is 50 metres across, that's the measure of the diameter, all the way across. So the radius is half that or 25 metres. Using the formula to find the area,

$$A = \pi r^2 = \pi \cdot 25^2 = 3.14 \cdot 625 = 1{,}962.5 \text{ square metres}$$

Using area formulas for different shapes

The area formulas will come up again and again in your maths journey, and Figure 15-5 outlines the most important (and useful) ones to remember. As you move into more senior years, the more complex formulas will likely be provided to you on a formula sheet when you are sitting a test or an exam. However, learning and memorising as many as you can is still a great idea, because doing so will save the most important thing in a test or exam environment — time!

Seeing is remembering, and writing is also remembering. Make charts of the main area formulas and plaster them anywhere that you spend lots of time — near your desk, on your bedside table, the back of the toilet door and even the outside of the shower screen. Make some cards containing these important facts and pop them in your pencil case, and look at them when you have a moment on the train or on the bus. They will soon be second nature to you.

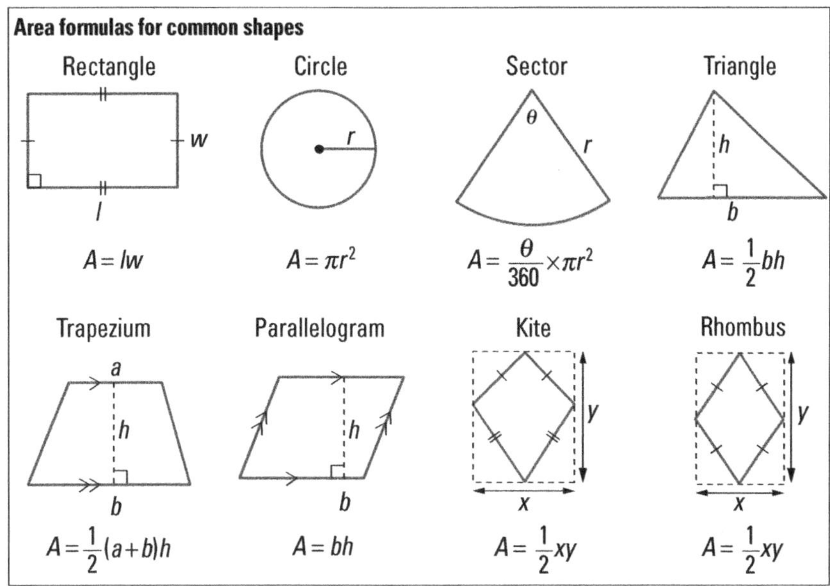

Figure 15-5: The area formulas.

Also keep in mind the golden rules to use when substituting values into area formulas (which also apply for any substitution or formula use):

- ✔ Identify the shape (or what you are trying to find)
- ✔ Draw a simple diagram
- ✔ Write down the rule
- ✔ Fill in what you know
- ✔ Solve
- ✔ Write your answer clearly

For example, say you need to calculate the area of the sector shown in Figure 15-6, and write your answer correct to two decimal places.

Write the formula and a list of what you need to know.

$$A = \frac{\theta}{360} \times \pi r^2$$

$$\theta = \frac{3}{4} \times 360$$

$$\theta = 270°$$

$$r = 15 \text{ cm}$$

Fill in what you know.

$$A = \frac{270}{360} \times \pi \times 15^2$$
$$= 530.143\,760\ldots$$

Round your answer to two decimal places and write your answer. Remember to include the appropriate units.

$$A = 530.14 \text{ cm}^2$$

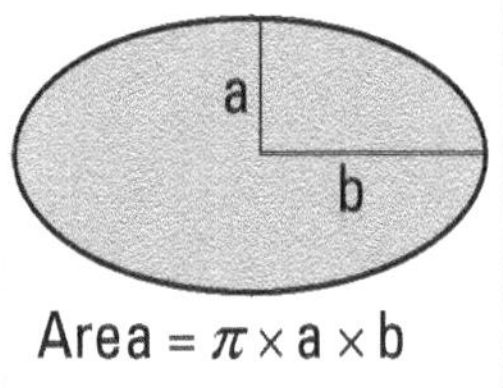

Figure 15-6: Calculating the area of an ellipse.

Working with composite shapes

Sometimes the shape provided is made up of two shapes. In order to calculate the total area, you need to divide the larger shape into two, three or even more smaller shapes that you recognise. You can then calculate the area of each part and find the total of all of the individual parts.

Sometimes the area required is shaded.

For example, say you need to calculate the area of the shaded area shown in Figure 15-7. You need to calculate the area of the larger circle (radius = 10 cm) and the smaller circle (radius = 8 cm) and then subtract the smaller area from the larger area.

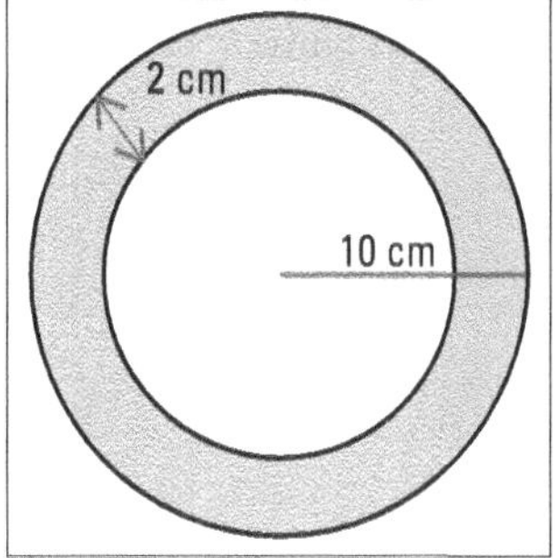

Figure 15-7: Calculating the area of a composite shapes.

Pumping Up with Volume Formulas

Area is a two-dimensional figure or representation. It's a flat region. Volume is three-dimensional. Unlike your last loser boyfriend or girlfriend, it has depth. To find volume, you measure across, front to back, and up and down.

With volume, you count how many cubes (picture sugar cubes) you can fit into an object. These cubes can be 1 centimetre on each edge, 1 metre on each edge, or however big they need to be. And, in keeping with the cube theme, you measure volume in cubic centimetres, cubic metres and cubic whatevers.

Prying into prisms and boxes

The volume of a rectangular prism, better known as a box, is one of the simplest to find in the world of volume problems. The bottom and top of a prism have exactly the same measurements. The distance from the top to bottom is the same, no matter where you measure, as long as you keep that distance perpendicular to both top and bottom.

The formula for finding the volume of a prism is $V = lwh$, which means that the volume is equal to the product of the length, l, times the width, w, times the height, h.

For example, find the volume of a box that is 4 metres long, 3 metres wide and 9 metres high.

$$V = lwh = (4)(3)(9) = 108 \text{ cubic metres}$$

That's 108 cubes, all 1 metre by 1 metre by 1 metre, that fit into the box.

Say your dad is looking at buying a 6-cubic-metre shed; what are the dimensions (how big is it)?

You can multiply three numbers together to get 6 in infinite ways. Go through some integers and some fractions.

Try to picture what the shed would look like with these dimensions.

- ✔ $6 = (1)(1)(6)$. That's 1 metre long, 1 metre wide, and 6 metres tall.

- ✔ $6 = (2)(1)(3)$. That's 2 metres long, 1 metre wide, and 3 metres tall.

- ✔ $6 = (1)(3)(2)$. That's 1 metre long, 3 metres wide, and 2 metres tall.

You could also try combinations of fractions. Which shed would you want? How tall are you? How much do you want to store at the back?

Cycling cylinders

Cylinders were my brother's favourite shape when he was in the navy. Being the wonderful sister that I am, I would send him chocolate chip cookies that fit exactly into a 1-kilogram coffee can. Imagine a stack of chocolate chip cookies coming to you every couple of weeks. Was he ever popular on *that* ship!

The formula for the volume of a cylinder is $V = \pi r^2 h$. The volume is equal to π times the radius (halfway across a circle) squared times the height.

A cylinder is a solid figure with a circle for a base. A can of tuna, a can of peas, a roll of toilet paper and, of course, a coffee can are all examples of cylinders. The tops and bottoms are circles, and the height of a cylinder is the distance between the circles.

To find the volume of a cylinder, you need the radius of the top and bottom, and you need the height. This formula tells you how many cubes will fit in the cylinder — like putting square pegs in a round hole, just trim them a bit.

For example, find the volume of an above-ground swimming pool that has a radius of 4 metres and a height of 1 metre.

Using the formula for the volume of a cylinder:

$$V = \pi r^2 h = \pi(4^2)(1) = \pi(16) \approx 3.14(16) \approx 50.24 \text{ cubic metres of water}$$

Scaling a pyramid

A pyramid is an easy thing to describe because you most likely have a mental picture of what a pyramid looks like. Technically, a pyramid is an object with a base (bottom) and triangles coming up from each side of the base to meet at a point.

The pyramids in Egypt have squares for the bottom and same-size triangles on the sides — at least, that's how they started. The wind and sand have eroded the tops so the Egyptian pyramids don't come to a point anymore. But the base of a pyramid can be an *equilateral triangle* (all three sides are the same length), a square, a *regular pentagon* (five sides, all the same length), and so on. The example shown here, however, sticks with square bases.

The formula for the volume of a pyramid is $V = \frac{1}{3}(\text{area of base}) \cdot h$.

Find the original volume of the Great Pyramid, which originally had a square base with each side measuring 756 feet and a height of 480 feet.

The base is a square, so the area of the base is s^2:

$$V = \frac{1}{3}s^2 \cdot h = \frac{1}{3}(756)^2 \cdot 480 = 91,445,760 \text{ cubic feet}$$

Pointing to cones

The formula for the volume of a cone is $V = \frac{1}{3}\pi r^2 h$.

The formula for finding the volume of a cone should look familiar for two reasons. First, it has the $\frac{1}{3}$, like the pyramid formula has. The one-third factor is common when a figure goes up into a single point. The other familiar part is the $\pi r^2 h$, which is the formula for finding the volume of a cylinder. You can think of a cone as being just a cylinder that was whittled away. The pointy-bottomed ice-cream cone is a classic cone shape, as are traffic pylons.

For example, what is the volume of a cone-shaped circus tent that has a diameter of 18 metres and a height of 20 metres?

If the diameter is 18 metres, then the radius is 9 metres:

$$V = \frac{1}{3}\pi r^2 h = \frac{1}{3}\pi(9)^2 \cdot 20 = 540\pi \approx 1,696 \text{ cubic feet}$$

Rolling along with spheres

A sphere is a familiar shape. Soccer balls, cricket balls, marbles and globes are all spheres. You need only one thing to find the volume of a sphere: The radius, which is the distance from the centre of the sphere to the outside.

The formula to determine the volume of a sphere is $V = \frac{4}{3}\pi r^3$.

For example, what is the volume of a ball that has a diameter of 18 centimetres?

A diameter of 18 centimetres means that the radius of the ball is 9 centimetres:

$$V = \frac{4}{3}\pi r^3 = \frac{4}{3}\pi \cdot 9^3 = 972\pi \approx 3{,}052$$

What is the volume of a sphere with a diameter of 4 centimetres?

$$V = \frac{4}{3}\pi r^3 = \frac{4}{3}\pi \cdot 2^3 = 10\frac{2}{3}\pi \approx 33.5$$

Triggering Trigonometric Ratios

You can use some fancy little ratios to work out missing lengths or angles in triangles — they're the trigonometric ratios, and this section tells you all about them.

Finding lengths

In some instances, you're given a right-angle triangle with the measurements for one length and one angle included. You can use this information to calculate the other lengths in the triangle using trigonometric ratios.

You have three sets of trigonometric ratios to remember. (As I advise when trying to remember anything, make a chart and stick it everywhere, and also write the formulas on easy to access cards.)

The ratios are sine, cosine and tangent, or sin, cos and tan, as follows.

$$\sin\theta = \frac{opposite}{hypotenuse} \qquad \cos\theta = \frac{adjacent}{hypotenuse} \qquad \tan\theta = \frac{opposite}{adjacent}$$

A really good way to remember these ratios is using the mnemonic *SOH, CAH, TOA*.

When using the ratios to find lengths, before doing anything else you need to draw a diagram and carefully label the sides of the triangle, remembering the hypotenuse is always directly opposite the right angle. The other two sides are labelled in reference to the given angle — one is opposite the angle and the other is adjacent to (next to) the angle. The two examples in Figure 15-8 show how these labels work.

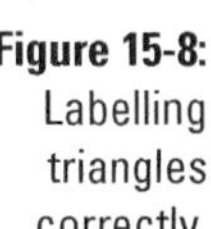

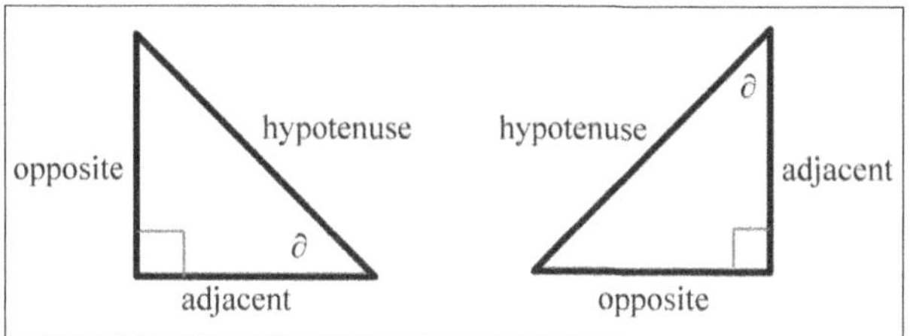

Figure 15-8: Labelling triangles correctly.

This is where my favourite rules again come in really handy:

- Identify the shape (or what you are trying to find)
- Draw a simple diagram
- Write down the rule
- Fill in what you know
- Solve
- Write your answer clearly

For example, say you're asked to find the length marked x in the triangle shown in Figure 15-9, and write your answer correct to two decimal places.

Look carefully at your labels and at the three trigonometric ratios. The one which uses the opposite and the hypotenuse is sine.

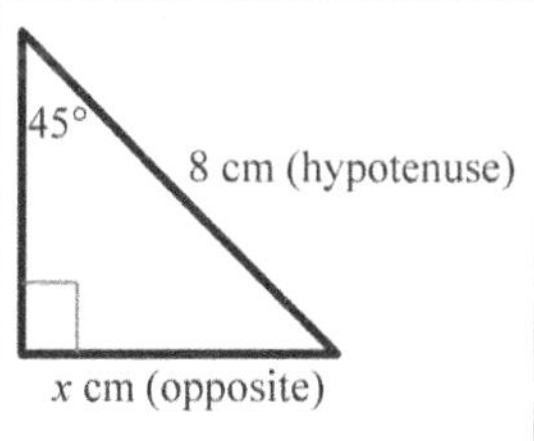

Figure 15-9: Using the correct ratio to find lengths.

Write the rule.

$$\sin\theta = \frac{opposite}{hypotenuse}$$

Fill in what you know.

$$\sin 45 = \frac{x}{8}$$

Now this looks just like solving an equation. You can move the divided by 8 to the other side by doing the inverse.

$$8 \times \sin 45 = x$$

If you prefer the variable on the left side, that's fine — just swap the sides. You can then use your calculator to evaluate and write your answer.

$$x = 8 \times \sin 45$$

$$x = 5.66 \text{ cm (correct to two decimal places)}$$

Figure 15-10 shows another example — this one ends a little differently.

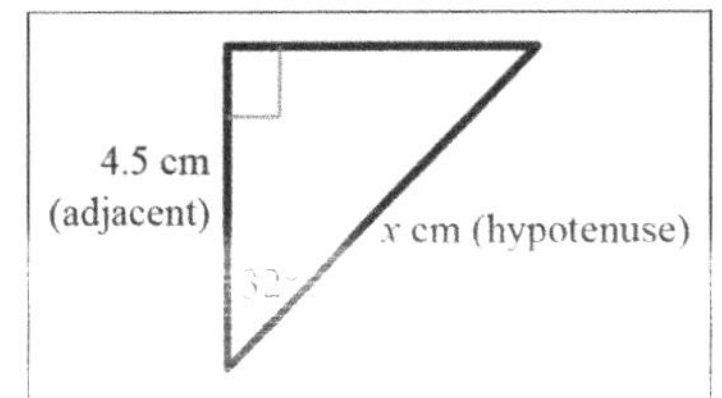

Figure 15-10: Finding x with a different ratio.

Look carefully at your labels and at the three trigonometric ratios. The one that uses the adjacent and the hypotenuse is cosine (cos).

Write the rule.

$$\cos \theta = \frac{adjacent}{hypotenuse}$$

Fill in what you know.

$$\cos 32 = \frac{4.5}{x}$$

Again, now this look just like solving an equation. In this case, however, the variable is on the bottom of the equation.

Move the divided by x to the other side by doing the inverse and then move the cos 32.

$$x \times \cos 32 = 4.5$$

$$x = \frac{4.5}{\cos 32}$$

So this time you need to divide the length to find the other length.

This will consistently be the case, as follows:

- ✔ If the x is on the bottom of the ratio the question requires multiplication.

- ✔ If the x is on the top of the ratio the question requires division.

Use your calculator to evaluate and write your answer.

$$x = \frac{4.5}{\cos 32}$$

$$x = 5.31 \text{ cm (correct to two decimal places)}$$

Finding angles

Sometimes all of the length information is given and you are required to find the size of the angle. The same process applies (see the preceding section for a breakdown of this process).

For example, find the angle $x°$ in Figure 15-11 and write your answer to the nearest whole decimal.

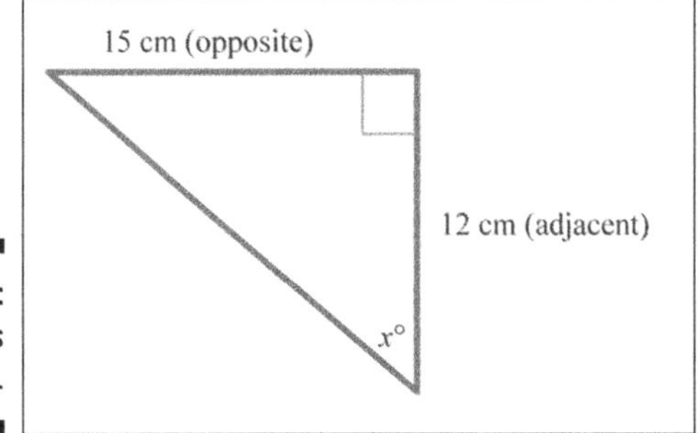

Figure 15-11:
Finding x as
an angle.

Look carefully at your labels and at the three trigonometric ratios. The one that uses the opposite and the adjacent is tangent (tan).

Write the rule.

$$\tan \theta = \frac{opposite}{adjacent}$$

Fill in what you know.

$$\tan x = \frac{15}{12}$$

Finding the angle is like doing the reverse of finding a length (refer to preceding section), so you need to utilise a special key on your calculator.

First, convert the fraction to a decimal.

$$\tan x = 1.25$$

Now you need to find the inverse tan of 1.25. On your calculator press the 2ND or CONTROL key and then the tan key. This should appear on your screen.

$$\tan^{-1} ($$

Type the decimal equivalent of your fraction (in this example, 1.25) into your calculator and press the = key.

$$\tan^{-1} (1.25) = 51.34°$$

Round your answer to the nearest whole degree and make sure that your answer is clear.

$$x = 52°$$

Understanding degrees and minutes

You may have noticed in the final example in the preceding section that you calculated the angle and got a decimal answer. In the example, the answer was required to be rounded to the nearest whole degree, which in this case meant it went to 52°, rather than 51°. Perhaps you're wondering why.

Angles are measured in degrees, with every degree equal to 60 minutes. And every minute is equal to 60 seconds. Because 34 is more than half of 60, you round up. (Other measures are also used in Years 11 and 12, but for now you will probably only need to calculate angles to degrees and minutes, so I will concentrate on that.)

With the answer calculated in the previous example, $\tan^{-1}(1.25) = 51.34°$.

If you were required to convert the decimal portion of the answer to minutes, you'd need to multiply 0.34 by 60.

$$0.34 \times 60 = 20.4'$$

If you were required to write your answer correct to the nearest minute, it would be $51°20'$.

Chapter 16

Geometry Basics

In This Chapter

▷ Starting out with lines, segments and rays

▷ Working out your angle

▷ Tricking out with triangles

▷ Finding similarities between different figures

*H*ave you ever reflected on the fact that you're literally surrounded by shapes? Look around. The rays of the sun are — what else? — rays. The book in your hands has a shape, every table and chair has a shape, every wall has an area, and every container has a shape and a volume; most picture frames are rectangles, DVDs are circles, soup cans are cylinders, and so on.

In this chapter, I take you through geometric shapes and proofs, looking at angles, triangles and similar shapes and figures.

Geometry Proofs

A *geometry proof* — like any mathematical proof — is an argument that begins with known facts, proceeds from there through a series of logical deductions, and ends with the thing you're trying to prove. Here's a very simple example using the line segments in Figure 16-1.

Figure 16-1:
PS and *WZ*,
each made
up of three
pieces
(*PQ* +
QR + *RS*
and *WX* +
XY + *YZ*).

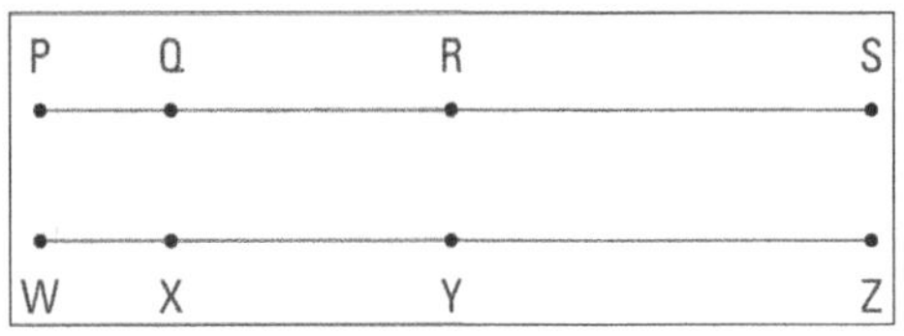

For this proof, you're told that segment $\overline{PS}$ is *congruent to* (the same length as) segment $\overline{WZ}$, that $\overline{PQ}$ is congruent to $\overline{WX}$, and that $\overline{QR}$ is congruent to $\overline{XY}$. You have to prove that $\overline{RS}$ is congruent to $\overline{YZ}$.

Now, you may be thinking, That's obvious — if $\overline{PQ}$ is the same length as $\overline{WZ}$ and both segments contain these equal short pieces and the equal medium pieces, then the longer third pieces have to be equal as well. And you'd be right. But that's not how the proof game is played. You have to spell out every little step in your thinking. Here's the whole chain of logical deductions:

1. $\overline{PQ}$ is congruent to $\overline{WZ}$ (this is given).

2. $\overline{PQ}$ and $\overline{WX}$ (these facts are also given).

3. Therefore, $\overline{PS}$ is the same length as $\overline{WZ}$ (because if you add equal things to equal things, you get equal totals).

4. As $\overline{PQ}$ is congruent to $\overline{WZ}$, segments $\overline{PQ}$ and $\overline{WX}$ are equal, and $\overline{QR}$ and $\overline{XY}$ are equal it would follow that $\overline{RS}$ is the same length as $\overline{YZ}$.

Am I Ever Going to Use This?

You'll likely have plenty of opportunities to use your knowledge about the geometry of shapes. What about geometry proofs? Not so much.

When you'll use your knowledge of shapes

Shapes are everywhere, so every educated person should have a working knowledge of shapes and their properties. In the future, if you have to buy fertiliser or grass seed for your lawn, you should know something about area. You might want to understand the volume measurements in cooking recipes, or you may want to work on an art or science project that involves geometry. You'll certainly need to understand something about geometry to build some shelves or a backyard deck. And after finishing your work, you might be hungry — a grasp of how area works can come in handy when you're ordering pizza: A 20-centimetre pizza is four, not two, times as big as a 10-centimetre one. There's no end to the list of geometry problems that come up in everyday life.

When you'll use your knowledge of proofs

Will you ever use your knowledge of geometry proofs? I'll give you a politically correct answer and a politically incorrect one. Take your pick.

First, the politically correct answer (which is also *actually* correct). Granted, it's extremely unlikely that you'll ever have occasion to do a single geometry proof outside of a high school maths course. However, doing geometry proofs teaches you important lessons that you can apply to nonmathematical arguments. Proofs teach you

- Not to assume things are true just because they seem true.
- To carefully explain each step in an argument even if you think it should be obvious to everyone.
- To search for holes in your arguments.
- Not to jump to conclusions.

In general, proofs teach you to be disciplined and rigorous in your thinking and in communicating your thoughts.

If you don't buy that PC stuff, I'm sure you'll get this politically incorrect answer: Okay, so you're never going to use geometry proofs, but you want to get a decent grade in maths, right? So you might as well pay attention in class (what else is there to do, anyway?), do your homework, and use the hints, tips and strategies I give you in this book. They'll make your life much easier. Promise.

Getting Down with Definitions

The study of geometry begins with the definitions of the five simplest geometric objects: Point, line, segment, ray and angle. And I throw in two extra definitions for you (plane and 3-D space) for no extra charge.

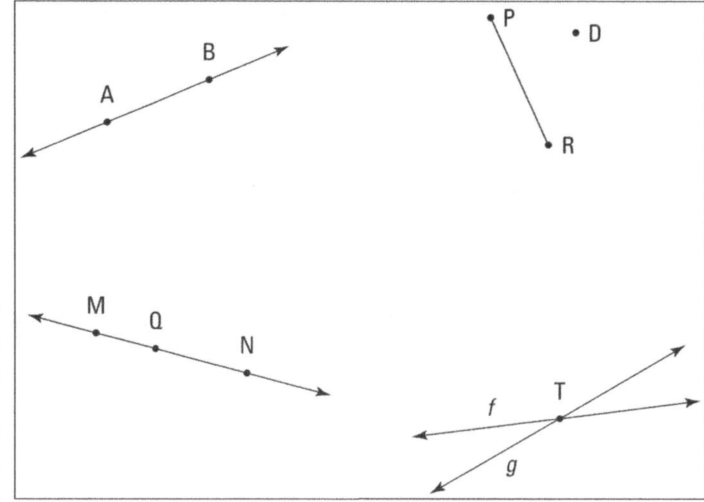

Figure 16-2: Some points, lines and segments.

✔ **Point:** A point is like a dot except that it has no size at all. A point is zero-dimensional, with no height, length or width, but you draw it as a dot anyway. You name a point with a single uppercase letter, as with points *A, D* and *T* in Figure 16-2.

✔ **Line:** A line is like a thin, straight wire (although really it's infinitely thin — or better yet, it has no width at all). Lines have length, so they're one-dimensional. Remember that a line goes on forever in both directions, which is why you use the little double-headed arrow as in $\overleftrightarrow{AB}$ (read as *line AB*).

Check out Figure 16-2 again. Lines are usually named using any two points on the line, with the letters in any order. Lines in Figure 16-2 are $\overleftrightarrow{AB}$ or $\overleftrightarrow{MN}$. Occasionally, lines are named with a single, italicised, lowercase letter, such as lines *f* and *g*.

✔ **Line segment (or just segment):** A segment is a section of a line that has two endpoints. See Figure 16-2 yet again. If a segment goes from *P* to *R*, you call it *segment PR* and write it as $\overline{PR}$. You can also switch the order of the letters and call it $\overline{RP}$. Segments can also appear within lines, as in $\overline{MN}$.

✔ **Ray:** A ray is a section of a line (kind of like half a line) that has one endpoint and goes on forever in the other direction. If its endpoint is point *K* and it goes through point *S* and then past *S* forever, you call it *ray KS* and write $\overrightarrow{KS}$. See Figure 16-3. Note that a ray like $\overrightarrow{AC}$ can also be called $\overrightarrow{AB}$ because either way, you start at *A* and go forever past *B* and *C*.

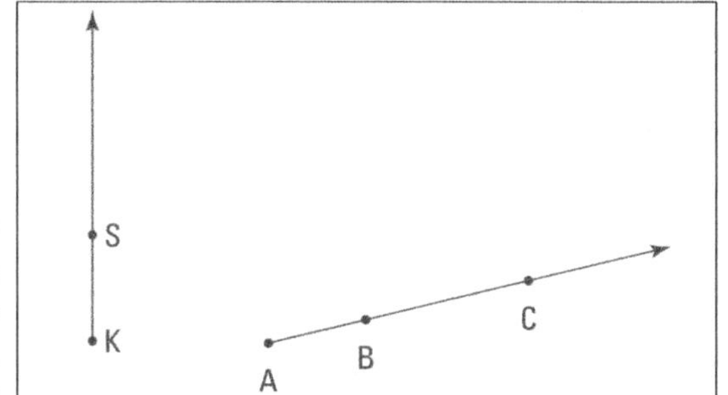

Figure 16-3: Catching a few rays.

- ✔ **Angle:** Two rays with the same endpoint form an angle. Each ray is a *side* of the angle, and the common endpoint is the angle's *vertex*. You can name an angle using its vertex alone or three points (first, a point on one ray, then the vertex, and then a point on the other ray).

  Check out Figure 16-4. Rays $\overline{PQ} \cong \overline{WX}$ and $\overline{QR} \cong \overline{XY}$ form the sides of an angle, with point P as the vertex. You can call the angle $\angle P$, $\angle RPQ$ or $\angle QPR$. Angles can also be named with numbers, such as the angle on the right in the figure, which you can call $\angle 4$. The number is just another way of naming the angle, and it has nothing to do with the size of the angle. The angle on the right also illustrates the *interior* and *exterior* of an angle.

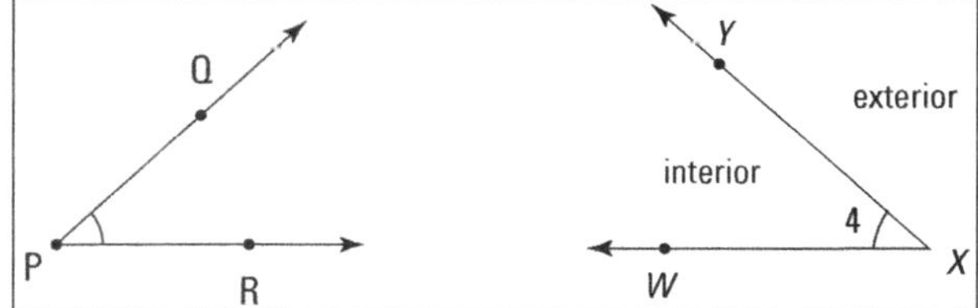

Figure 16-4: Some angles and their parts.

- ✔ **Plane:** A plane is like a perfectly flat sheet of paper except that it has no thickness whatsoever and it goes on forever in all directions. You might say it's infinitely thin and has an infinite length and an infinite width. Because it has length and width but no height, it's two-dimensional. Planes are named with a single, italicised, lowercase letter or sometimes with the name of a figure (a rectangle, for example) that lies in the plane.

- ✔ **3-D (three-dimensional) space:** 3-D space is everywhere — all of space in every direction. First, picture an infinitely big map that goes forever to the north, south, east and west. That's a two-dimensional plane. Then, to get 3-D space from this map, add the third dimension by going up and down forever.

A Few Points on Points

Not much can be said about points. They have no features, and each one is the same as every other. Various *groups* of points, however, do merit an explanation:

- **Collinear points:** See the word *line* in *collinear*? Collinear points are points that lie on a line. Any two points are always collinear because you can always connect them with a straight line. Three or more points can be collinear, but they don't have to be.

- **Non-collinear points:** Non-collinear points are three or more points that don't all lie on the same line.

- **Coplanar points:** A group of points that lie in the same plane are coplanar. Any two or three points are always coplanar. Four or more points might or might not be coplanar.

  Look at Figure 16-5, which shows coplanar points *A*, *B*, *C* and *D*. The box on the right has many sets of coplanar points. Points *P*, *Q*, *X* and *W*, for example, are coplanar; the plane that contains them is the left side of the box. Note that points *Q*, *X*, *S* and *Z* are also coplanar even though the plane that contains them isn't shown; it slices the box in half diagonally.

- **Non-coplanar points:** A group of points that don't all lie in the same plane are non-coplanar. In Figure 16-5, points *P*, *Q*, *X* and *Y* are non-coplanar. The top of the box contains *Q*, *X* and *Y*, and the left side contains *P*, *Q* and *X*, but no plane contains all four points.

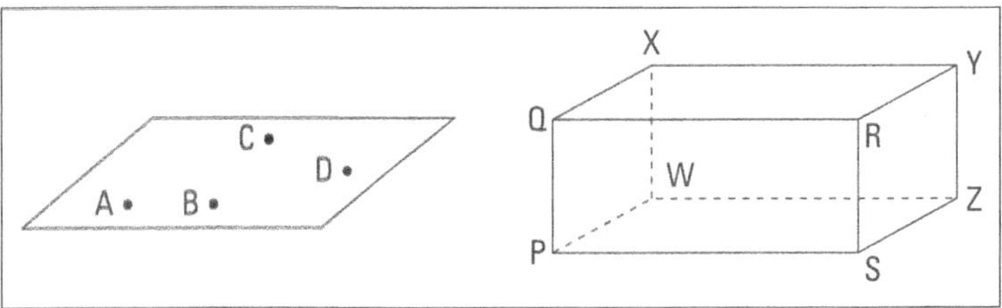

Figure 16-5: Coplanar and non-coplanar points.

Lines, Segments and Rays

In this section, I describe different types of lines (or segments or rays) or pairs of lines (or segments or rays) based on the direction they're pointing or how they relate to each other.

Horizontal and vertical lines

Defining *horizontal* and *vertical* may seem a bit pointless because you probably already know what the terms mean. But, hey, this is a maths book, and maths books are supposed to define terms. Who am I to question this tradition?

- ✔ **Horizontal lines, segments or rays:** Horizontal lines, segments and rays go straight across, left and right, not up or down at all — you know, like the horizon.

- ✔ **Vertical lines, segments or rays:** Lines or parts of a line that go straight up and down are vertical. (Shocking!)

Doubling up with pairs of lines

In this section, I give you five terms that describe pairs of lines. The first four are about coplanar lines — you use these a lot. The fifth term describes non-coplanar lines. This term comes up only in 3-D problems, so you won't need it much.

Coplanar lines

Coplanar lines are lines in the same plane. Here are some ways coplanar lines may interact:

- ✔ **Parallel lines, segments or rays:** Lines that run in the same direction and never cross (like two railroad tracks) are called parallel. Segments and rays are parallel if the lines that contain them are parallel. If $\overline{PR} \cong \overline{WY}$ is parallel to $\overline{RS} \cong \overline{YZ}$, you write $\overline{PS}$.

- ✔ **Intersecting lines, segments or rays:** Lines, rays or segments that cross or touch are intersecting.

- ✔ **Perpendicular lines, segments or rays:** Lines, segments or rays that intersect at right angles are perpendicular. If $\overline{PQ}$ is perpendicular to, $\overline{RS}$ you write $\overline{PQ} \perp \overline{RS}$. See Figure 16-6. The little boxes in the corners of the angles indicate right angles.

- ✔ **Oblique lines, segments or rays:** Lines or segments or rays that intersect at any angle other than 90° are called *oblique*. See Figure 16-6.

Non-coplanar lines

Non-coplanar lines are lines that cannot be contained in a single plane.

Lines that don't lie in the same plane are called skew lines — *skew* simply means *non-coplanar*. Or you can say that skew lines are lines that are neither parallel nor intersecting.

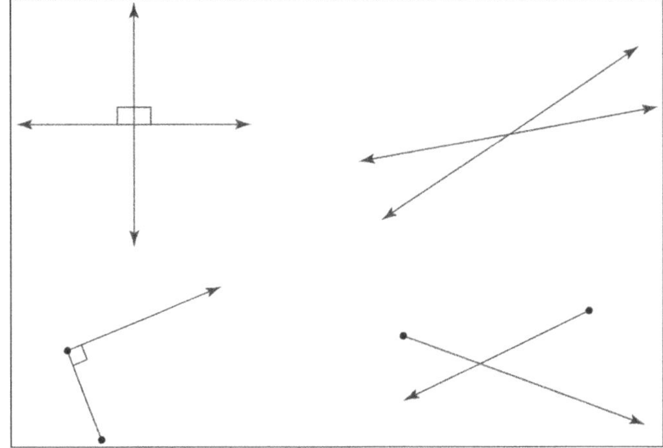

Figure 16-6:
Perpendicular and oblique lines, rays and segments.

Investigating the Plane Facts

Here are two terms for a pair of planes (see Figure 16-7):

- **Parallel planes:** Parallel planes are planes that never cross. The ceiling of a room (assuming it's flat) and the floor are parallel planes (though true planes extend forever).

- **Intersecting planes:** Hold onto your hat — intersecting planes are planes that cross or intersect. When planes intersect, the place where they cross forms a line. The floor and a wall of a room are intersecting planes, and where the floor meets the wall is the line of intersection of the two planes.

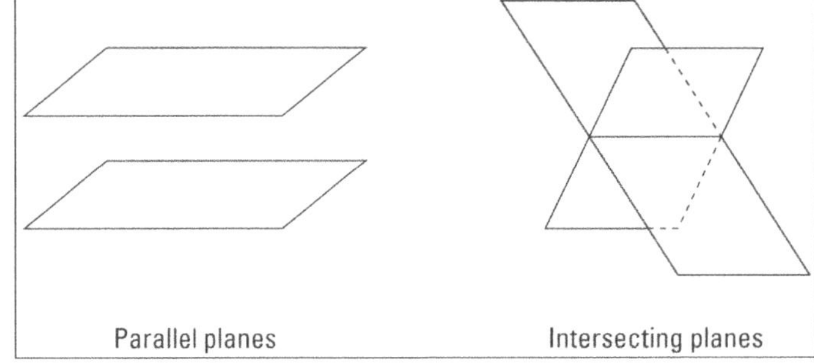

Figure 16-7:
Parallel and intersecting planes.

Everybody's Got an Angle

Angles are one of the basic building blocks of triangles and other polygons. Angles appear on virtually every page of every geometry book, so you gotta get up to speed about them — no ifs, ands, or buts.

Five types of angles

Check out the five angle definitions and see Figure 16-8:

- **Acute angle:** An acute angle is less than 90°. Think 'a-*cute* little angle.'

- **Right angle:** A right angle is a 90° angle. Right angles should be familiar to you from the corners of picture frames, tabletops, boxes and books, and all kinds of other things that show up in everyday life.

- **Obtuse angle:** An obtuse angle has a measure greater than 90°.

- **Straight angle:** A straight angle has a measure of 180°; it looks just like a line with a point on it.

- **Reflex angle:** A reflex angle has a measure of more than 180°. Basically, a reflex angle is just the other side of an ordinary angle.

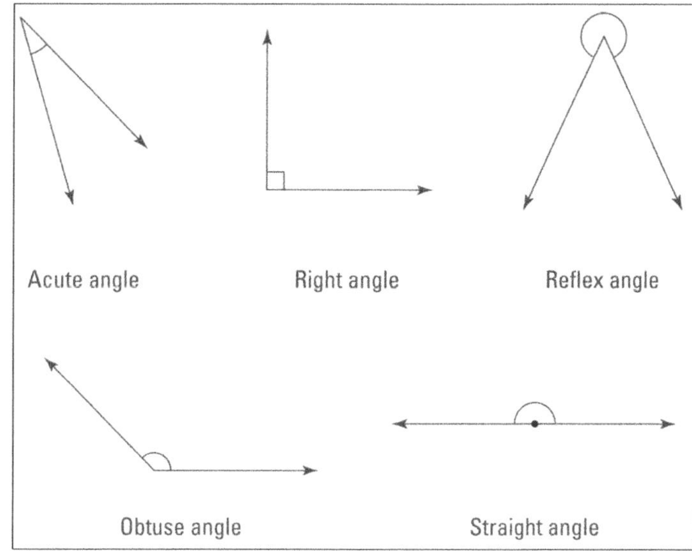

Figure 16-8:
Examining
all the
angles.

Angle pairs

The needy angles in this section have to be in a relationship with another angle for these definitions to mean anything:

- **Adjacent angles:** Adjacent angles are neighbouring angles that have the same vertex and that share a side; also, neither angle can be inside the other. I realise that's quite a mouthful. This very simple idea is kind of a pain to define, so just check out Figure 16-9. $\angle BAC$ and $\angle CAD$ are adjacent, as are $\angle 1$ and $\angle 2$. However, neither $\angle 1$ nor $\angle 2$ is adjacent to $\angle XYZ$ because they're both inside $\angle XYZ$. None of the unnamed angles to the right are adjacent because they either don't share a vertex or don't share a side.

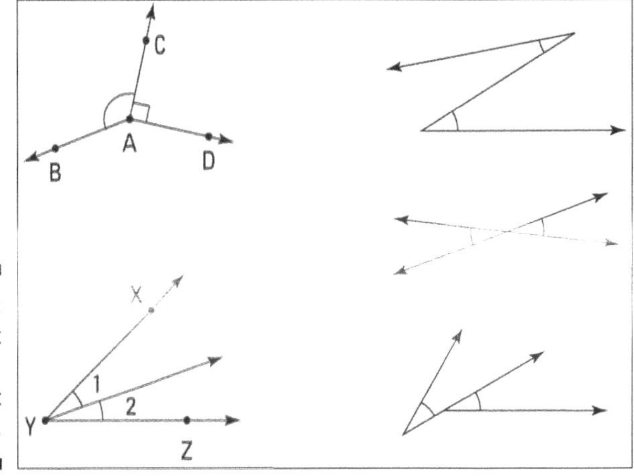

Figure 16-9: Adjacent and non-adjacent angles.

- **Complementary angles:** Two angles that add up to 90° are complementary. They can be adjacent angles but don't have to be. In Figure 16-10, adjacent angles $\angle 1$ and $\angle 2$ are complementary because they make a right angle; $\angle P$ and $\angle Q$ are complementary because they add up to 90°.

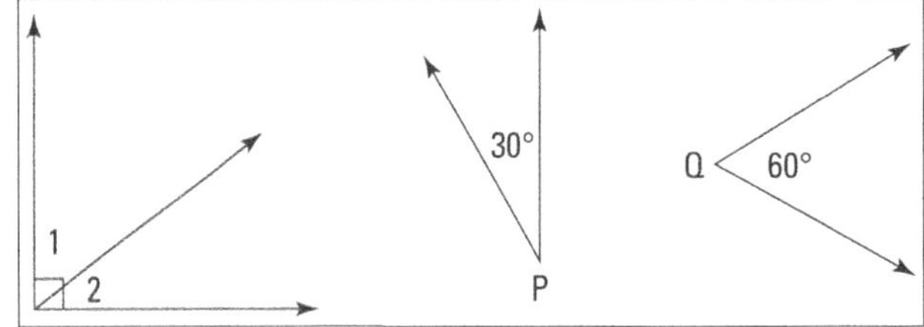

Figure 16-10: Complementary angles can join forces to form a right angle.

✔ **Supplementary angles:** Two angles that add up to 180° are supplementary. They may or may not be adjacent angles. In Figure 16-11, ∠1 and ∠2, or the two right angles, are supplementary because they form a straight angle. Such angle pairs are called a *linear pair*. Angles A and Z are supplementary because they add up to 180°.

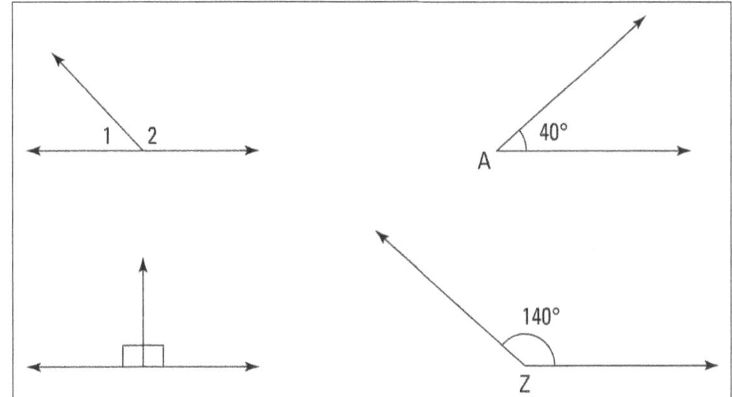

Figure 16-11: Together, supplementary angles can form a straight line.

✔ **Vertical angles:** When two intersecting lines form an X, the angles on the opposite sides of the X are called vertical angles. Two vertical angles are always congruent. By the way, the *vertical* in *vertical angles* has nothing to do with the up-and-down meaning of the word vertical.

Bisection and Trisection

For all you fans of bicycles and tricycles and bifocals and trifocals — not to mention the biathlon and the triathlon, bifurcation and trifurcation, and bipartition and tripartition — you're really going to love this section on bisection and trisection: Cutting something into two or three equal parts.

Segments

Segment *bisection,* the related term *midpoint,* and segment *trisection* are pretty simple ideas.

✔ **Segment bisection:** A point, segment, ray or line that divides a segment into two congruent segments *bisects* the segment.

✔ **Midpoint:** The point where a segment is bisected is called the *midpoint*; the midpoint cuts the segment into two congruent parts.

✔ **Segment trisection:** Two things (points, segments, rays or lines) that divide a segment into three congruent segments *trisect* the segment.

Students often make the mistake of thinking that *divide* means to *bisect*, or cut exactly in half. This error is understandable because when you do ordinary division with numbers, you are, in a sense, dividing the larger number into equal parts (24 ÷ 2 = 12 because 12 + 12 = 24). But in geometry, to *divide* something just means to cut it into parts of any size, equal or unequal. *Bisect* and *trisect*, of course, *do* mean to cut into exactly equal parts.

Angles

Brace yourself for a shocker: The terms *bisecting* and *trisecting* mean the same thing for angles as they do for segments! As follows:

✔ **Angle bisection:** A ray that cuts an angle into two congruent angles *bisects* the angle. The ray is called the *angle bisector*.

✔ **Angle trisection:** Two rays that divide an angle into three congruent angles *trisect* the angle. These rays are called *angle trisectors*.

Take a stab at this problem: In Figure 16-12, $\overline{TP}$ bisects $\angle STL$, which equals $(12x - 24)°$; $\overline{TL}$ bisects $\angle PTI$, which equals $(8x)°$. Is $\angle STI$ trisected, and what is its measure?

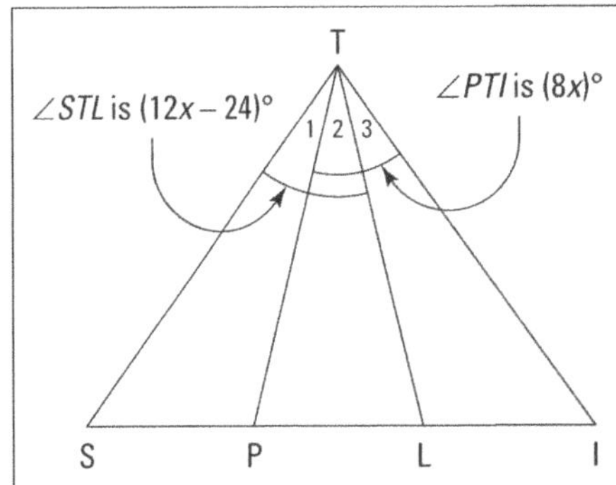

Figure 16-12:
A three-way
SPLIT.

Nothing to it. First, yes, $\angle STI$ is trisected. You know this because $\angle STL$ is bisected, so $\angle 1$ must equal $\angle 2$. And because $\angle PTI$ is bisected, $\angle 2$ equals $\angle 3$. Thus, all three angles must be equal, and that means $\angle STI$ is trisected.

Now find the measure of $\angle STI$. Because $\angle STL$ — which measures $(12x - 24)°$ — is bisected, $\angle 2$ must be half its size, or $(6x - 12)°$. And because $\angle PTI$ is bisected, $\angle 2$ must also be half the size of $\angle PTI$ — that's half of $(8x)°$, or $(4x)°$. Because $\angle 2$ equals both $(6x - 12)°$ and $(4x)°$, you set those expressions equal to each other and solve for x:

$$6x - 12 = 4x$$
$$2x = 12$$
$$x = 6$$

Then just plug $x = 6$ into, say, $(4x)°$, which gives you 4×6, or $24°$ for $\angle 2$. Angle STI is three times that, or $72°$. That does it.

When rays trisect an angle of a triangle, the opposite side of the triangle is *never* trisected by these rays. (It might be close, but it's never exactly trisected.) In Figure 16-12, for instance, because $\angle STI$ is trisected, $\overline{SI}$ is definitely *not* trisected by points P and L. This warning also works in reverse. Assuming for the sake of argument that P and L *did* trisect $\overline{SI}$, you would know that $\angle STI$ is definitely *not* trisected.

Taking In a Triangle's Sides

Triangles are classified according to the length of their sides or the measure of their angles. These classifications come in threes, just like the sides and angles themselves.

The following are triangle classifications based on sides:

- **Scalene:** A triangle with no congruent sides
- **Isosceles:** A triangle with at least two congruent sides
- **Equilateral:** A triangle with three congruent sides

Scalene triangles

In addition to having three unequal sides, scalene triangles have three unequal angles. The shortest side is across from the smallest angle, the medium side is across from the medium angle and the longest side is across from the largest angle.

Isosceles triangles

An isosceles triangle has two (or three) equal sides and two (or three) equal angles. The equal sides are called *legs* and the third side is the *base*. The two angles touching the base (which are congruent) are called *base angles*. The angle between the two legs is called the *vertex angle*. See Figure 16-13.

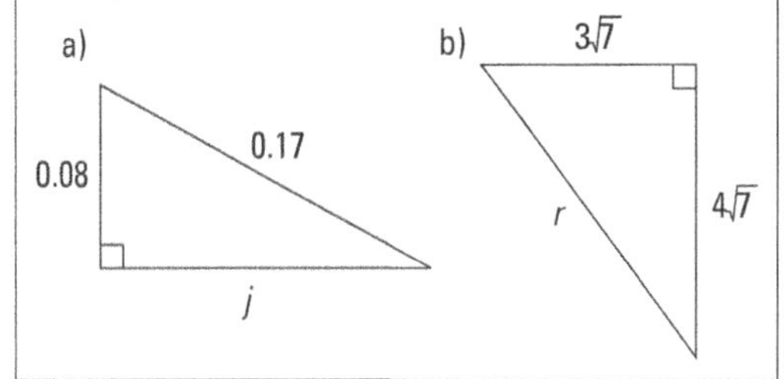

Figure 16-13: Two run-of-the-mill isosceles triangles.

Equilateral triangles

An equilateral triangle has three equal sides and three equal angles (which are each 60°). Its equal angles make it *equiangular* as well as equilateral. Note that an equilateral triangle is also isosceles.

Proving Triangles are Congruent

Before learning how to prove that triangles are congruent, you've got to know what congruent triangles are, right? Here you go . . .

Congruent triangles are triangles in which all pairs of corresponding sides and angles are congruent.

Maybe the best way to think about what it means for two triangles (or any other shapes) to be congruent is that you could move them around (by shifting, rotating and/or flipping them) so that they'd stack perfectly on top of one another. You indicate that triangles are congruent with a statement such as $\triangle ABC \cong \triangle XYZ$, which means that vertex A (the first letter) corresponds with and would stack on vertex X (the first letter), B would stack on Y and C would stack on Z. Side $\overline{AB}$ would stack on side $\overline{XY}$, $\angle B$ would stack on $\angle Y$, and so on.

So now, on to the methods for proving triangles congruent. There are five ways: SSS, SAS, ASA, AAS, and RHS.

SSS: The side-side-side method

If the three sides of one triangle are congruent to the three sides of another triangle, the triangles are congruent. Figure 16-14 illustrates this idea.

Figure 16-14:
Triangles with congruent sides are congruent.

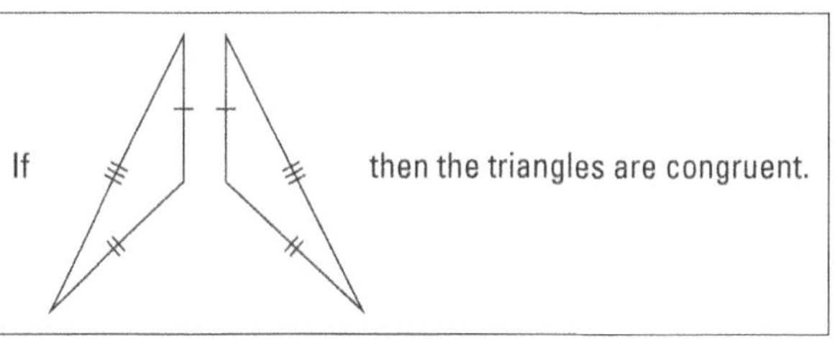

You can use SSS in the following '*TRIANGLE*' proof:

Given: $\overline{AG} \cong \overline{EG}$

$\overline{AG} \cong \overline{EG}$

$\overline{AR} \cong \overline{ET}$

$\overline{NI} \cong \overline{LI}$

T is the midpoint of $\overline{NI}$

R is the midpoint of $\overline{LI}$

Prove: $\triangle ANT \cong \triangle ELR$

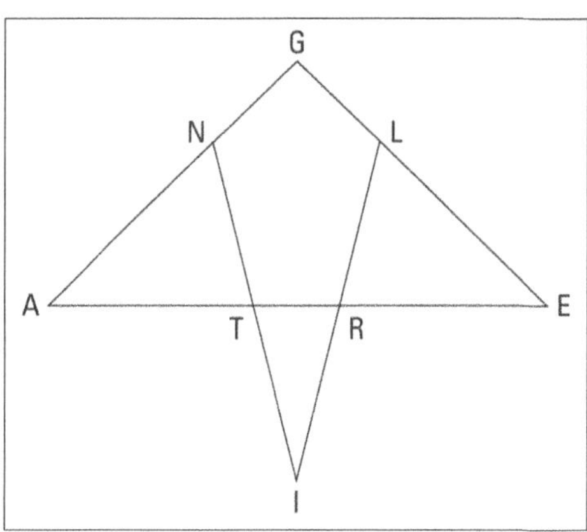

First, come up with a game plan. Here's how that might work.

You know you've got to prove the triangles congruent, so your first question should be, 'Can you show that the three pairs of corresponding sides are congruent?' Sure, you can do that:

- ✔ Subtract $\overline{NG}$ and $\overline{LG}$ from $\overline{AG}$ and $\overline{EG}$ to get the first pair of congruent sides, $\overline{AN}$ and $\overline{EL}$.

- ✔ Subtract $\overline{TR}$ from $\overline{AR}$ and $\overline{ET}$ to get the second pair of congruent sides, $\overline{AT}$ and $\overline{ER}$.

- ✔ Cut congruent segments $\overline{NI}$ and $\overline{LI}$ in half to get the third pair, $\overline{NT}$ and $\overline{LR}$. That's it.

To make the game plan more tangible, you may want to make up lengths for the various segments. For instance, say $\overline{AG}$ and $\overline{EG}$ are 9, $\overline{NG}$ and $\overline{LG}$ are 3, $\overline{AR}$ and $\overline{ET}$ are 8, $\overline{TR}$ is 3, and $\overline{NI}$ and $\overline{LI}$ are 8. When you do the maths, you see that $\triangle ANT$ and $\triangle ELR$ both end up with sides of 4, 5 and 6, which means, of course, that they're congruent.

Here's how the formal proof shapes up:

Statements	Reasons
1) $\overline{AG} \cong \overline{EG}$ $\overline{NG} \cong \overline{LG}$	1) Given.
2) $\overline{AN} \cong \overline{EL}$	2) If two congruent segments are subtracted from two other congruent segments, then the differences are congruent.
3) $\overline{AR} \cong \overline{ET}$	3) Given.
4) $\overline{AT} \cong \overline{ER}$	4) If a segment is subtracted from two congruent segments, then the differences are congruent.
5) $\overline{NI} \cong \overline{LI}$ T is the midpoint of $\overline{NI}$ R is the midpoint of $\overline{LI}$	5) Given.
6) $\overline{NT} \cong \overline{LR}$	6) If segments are congruent, then their Like Divisions are congruent (half of one equals half of the other).
7) $\triangle ANT \cong \triangle ELR$	7) SSS (2, 4, 6).

Note: After SSS in the final step, I indicate the three lines from the statement column where I've shown the three pairs of sides to be congruent. You don't have to do this, but it's a good idea. It can help you avoid some careless mistakes. Remember that each of the three lines you list must show a *congruence* of segments (or angles, if you're using one of the other approaches to proving triangles congruent).

SAS: side-angle-side

If two sides and the included angle of one triangle are congruent to two sides and the included angle of another triangle, the triangles are congruent. (The *included angle* is the angle formed by the two sides.) Figure 16-15 illustrates this method.

Figure 16-15:
Two sides and the angle between them make these triangles congruent.

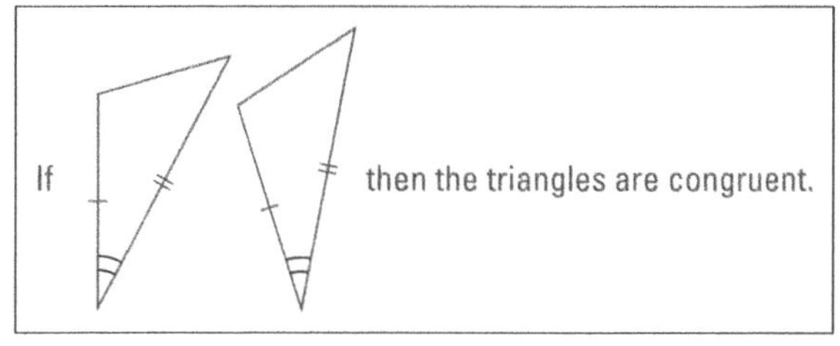

Check out the SAS postulate in action:

Given: $\triangle QZX$ is isosceles with base $\overline{QX}$

$\overline{JQ} \cong \overline{XF}$

$\angle 1 \cong \angle 2$

Prove: $\triangle JZX \cong \triangle FZQ$

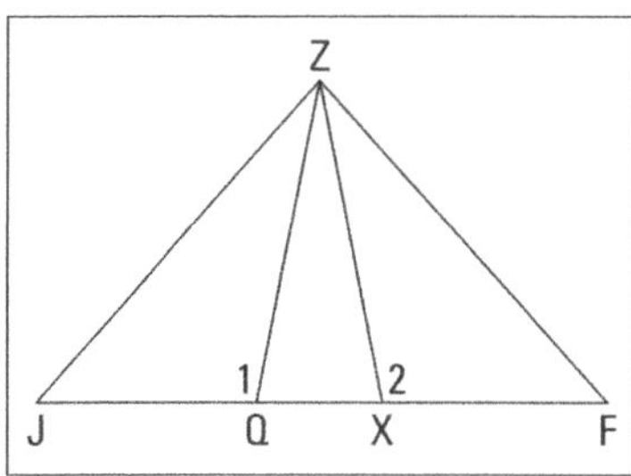

When overlapping triangles muddy your understanding of a proof diagram, try redrawing the diagram with the triangles separated. Doing so can give you a clearer idea of how the triangles' sides and angles relate to each other. Focusing on your new diagram may make it easier to figure out what you need to prove the triangles congruent. However, you still need to use the original diagram to understand some parts of the proof, so use the second diagram as a sort of aid to get a better handle on the original diagram.

Figure 16-16 shows you what this proof diagram looks like with the triangles separated.

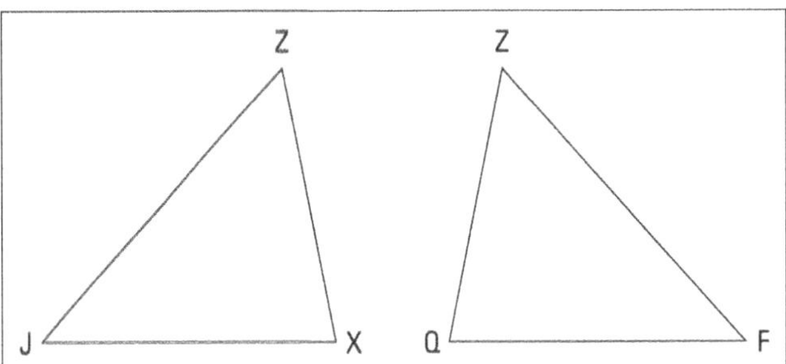

Figure 16-16:
An amicable separation of triangles.

Looking at Figure 16-16, you can easily see that the triangles are congruent (they're mirror images of each other). You also see that, for example, side $\overline{ZX}$ corresponds to side $\overline{ZQ}$ and that $\angle X$ corresponds to $\angle Q$.

So using both diagrams, here's a possible game plan:

- ✔ **Determine which congruent triangle postulate is likely to be the ticket for proving the triangles congruent.** You know you have to prove the triangles congruent, and one of the givens is about angles, so SAS looks like a better candidate than SSS for the final reason. (You don't have to figure this out now, but it's not a bad idea to at least have a guess about the final reason.)

- ✔ **Look at the givens and think about what they tell you about the triangles.** Triangle QZF is isosceles, so that tells you $\overline{ZQ} \cong \overline{ZX}$. Look at these sides in both figures. Put tick marks on $\overline{ZQ}$ and $\overline{ZX}$ in Figure 16-16 to show that you know they're congruent. Now consider why they'd tell you the next given, $\overline{JX} \cong \overline{QF}$. Well, what if they were both 6 and $\overline{QZ}$ were 2? $\overline{JX}$ and $\overline{QF}$ would both be 6, so you have a second pair of congruent sides. Put tick marks on Figure 16-16 to show this congruence.

✔ **Find the pair of congruent angles.** Look at Figure 16-16 again. If you can show that $\angle X$ is congruent to $\angle Q$, you'll have SAS. Do you see where $\angle X$ and $\angle Q$ fit into the original diagram? Note that they're the supplements of $\angle 1$ and $\angle 2$. That does it. Angles 1 and 2 are congruent, so their supplements are congruent as well. (If you fill in numbers, you can see that if $\angle 1$ and $\angle 2$ are both $100°$, $\angle Q$ and $\angle X$ would both be $80°$.)

Here's the formal proof:

Statements	Reasons
1) $\triangle QZF$ is isosceles with base $\overline{QF}$	1) Given.
2) $\overline{ZX} \cong \overline{ZQ}$	2) Definition of isosceles triangle.
3) JX is congruent to QF	3) Given.
4) $\overline{JX} \cong \overline{FQ}$	4) If a segment is added to two congruent segments, then the sums are congruent.
5) Angle X is congruent to angle Q	5) Given.
6) $\angle ZXJ \cong \angle ZQF$	6) If two angles are supplementary to two other congruent angles, then they're congruent.
7) $\triangle JZX \cong \triangle FZQ$	7) SAS (2, 6, 4).

ASA: The angle-side-angle tack

If two angles and the included side of one triangle are congruent to two angles and the included side of another triangle, the triangles are congruent. See Figure 16-17.

Figure 16-17:
Two angles and their shared side make these triangles congruent.

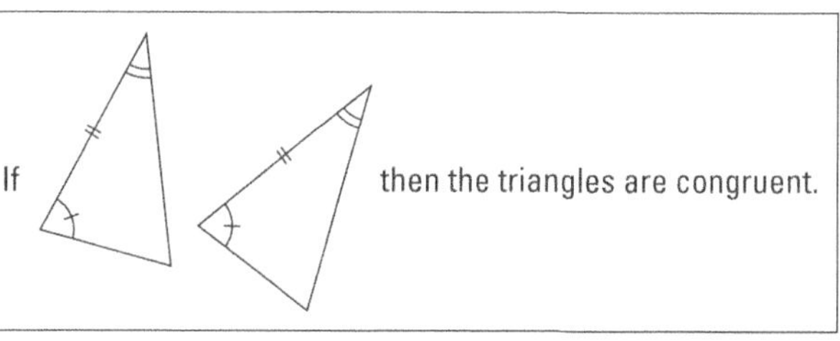

AAS: angle-angle-side

If two angles and a non-included side of one triangle are congruent to the corresponding parts of another triangle, the triangles are congruent. Figure 16-18 shows you how AAS works.

Figure 16-18: Two congruent angles and a side not between them make these triangles congruent.

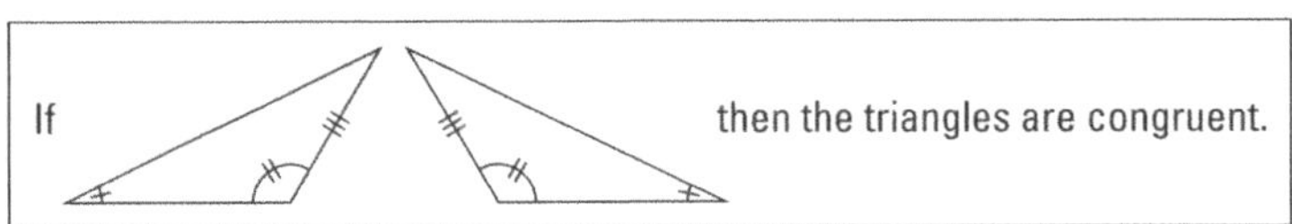

Like ASA, to use AAS, you need two pairs of congruent angles and one pair of congruent sides to prove two triangles congruent. But for AAS, the two angles and one side in each triangle must go in the order angle-angle-side (going around the triangle either clockwise or counter-clockwise).

ASS and SSA don't prove anything, so don't try using ASS (or its backward twin, SSA) to prove triangles congruent. You can use SSS, SAS, ASA and AAS (or SAA, the backward twin of AAS) to prove triangles congruent, but not ASS. In short, every three-letter combination of *A*s and *S*s proves something unless it spells *ass* or is *ass* backward. (You work with AAA in the later section in this chapter 'Proving Triangles Similar'.)

Last but not least: RHS

If the hypotenuse and a leg of one right-angled triangle are congruent to the hypotenuse and a leg of another right-angled triangle, the triangles are congruent. RHS is different from the other four ways of proving triangles congruent because it works only for right-angled triangles.

As with SSS, SAS, ASA and AAS — before you can use it in a proof, you need to have three things in the statement column (congruent hypotenuses, congruent legs and right angles).

Similar Figures

In this section, I cover the formal definition of similarity, how similar figures are named and how they're positioned.

Defining similar polygons

As you see in Figure 16-19, quadrilateral *WXYZ* is the same shape as quadrilateral *ABCD*, but it's ten times larger (though not drawn to scale). These quadrilaterals are therefore similar.

For two polygons to be similar, both of the following must be true:

- ✔ Corresponding angles are congruent.
- ✔ Corresponding sides are proportional.

To fully understand this definition, you have to know what *corresponding angles* and *corresponding sides* mean. Here's the lowdown on *corresponding*. In Figure 16-19, if you expand *ABCD* to the same size as *WXYZ* and slide it to the right, it'd stack perfectly on top of *WXYZ*. *A* would stack on *W*, *B* on *X*, *C* on *Y* and *D* on *Z*. These vertices are thus *corresponding*. And, therefore, you say that $\angle A$ corresponds to $\angle W$, $\angle B$ corresponds to $\angle X$ and so on. Also, side $\overline{AB}$ corresponds to side $\overline{WX}$, $\overline{BC}$ to $\overline{XY}$, and so on.

When you name similar polygons, pay attention to how the vertices pair up. For the quadrilaterals in Figure 16-19, you write that *ABCD* ~ *WXYZ* (the squiggle symbol means *is similar to*) because *A* and *W* (the first letters) are corresponding vertices, *B* and *X* (the second letters) are corresponding and so on. You can also write *BCDA* ~ *XYZW* (because corresponding vertices pair up) but *not ABCD* ~ *YZWX*.

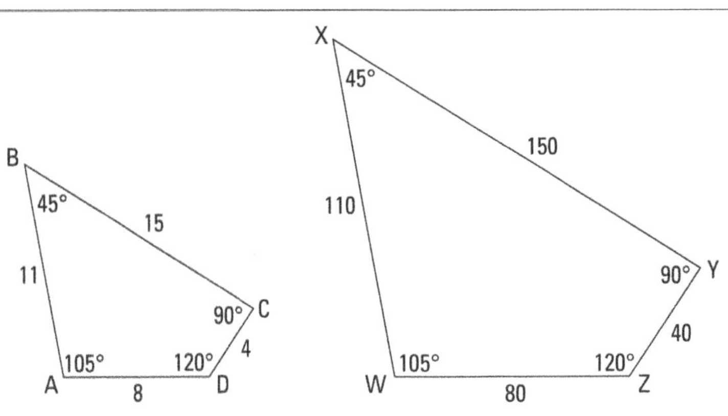

Now I'll use quadrilaterals *ABCD* and *WXYZ* to explore the definition of similar polygons in greater depth:

- **Corresponding angles are congruent.** You can see that ∠A and ∠W are both 105° and thus congruent, ∠B ≅ ∠X and so on. When you blow up or shrink a figure, the angles don't change.

- **Corresponding sides are proportional.** The ratios of corresponding sides are equal, like this:

$$\frac{\text{left side}_{WXYZ}}{\text{left side}_{ABCD}} = \frac{\text{top}_{WXYZ}}{\text{top}_{ABCD}} = \frac{\text{right side}_{WXYZ}}{\text{right side}_{ABCD}} = \frac{\text{base}_{WXYZ}}{\text{base}_{ABCD}}$$

$$\frac{WX}{AB} = \frac{XY}{BC} = \frac{YZ}{CD} = \frac{ZW}{DA}$$

$$\frac{110}{11} = \frac{150}{15} = \frac{40}{4} = \frac{80}{8} = 10$$

Each ratio equals 10, the expansion factor. (If the ratios were flipped upside down — which is equally valid — each would equal $\frac{1}{10}$, the shrink factor.) And not only do these ratios all equal 10, but the ratio of the perimeters of *ABCD* and *WXYZ* also equals 10.

How similar figures line up

Two similar figures can be positioned so that they either line up or don't line up. You can see that figures *ABCD* and *WXYZ* in Figure 16-19 are positioned in the same way in the sense that if you were to blow up *ABCD* to the size of *WXYZ* and then slide *ABCD* over, it'd match up perfectly with *WXYZ*. Now check out Figure 16-20, which shows *ABCD* again with another similar quadrilateral. You can easily see that, unlike the quadrilaterals in Figure 16-19, *ABCD* and *PQRS* are *not* positioned in the same way.

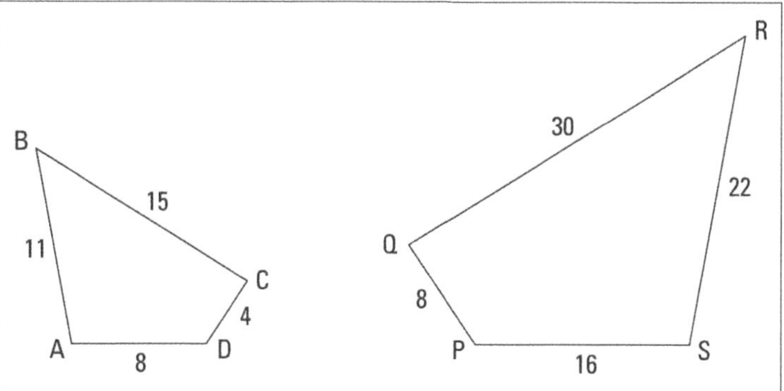

Figure 16-20:
Similar quadrilaterals that aren't lined up.

In the preceding section, you see how to set up a proportion for similar figures using the positions of their sides, which I've labelled *left side, right side, top* and *base* — for example, one valid proportion is $\frac{\text{left side}}{\text{left side}} = \frac{\text{top}}{\text{top}}$. This is a good way to think about how proportions work with similar figures, but this works only if the figures are drawn like *ABCD* and *WXYZ* are. When similar figures are drawn facing different ways, as in Figure 16-20, the left side doesn't necessarily correspond to the left side, and so on, and you have to take greater care that you're pairing up the proper vertices and sides.

Quadrilaterals *ABCD* and *PQRS* are similar, but you *can't* say that *ABCD* ~ *PQRS* because the vertices don't pair up in this order. Ignoring its size, *PQRS* is the mirror image of *ABCD*. If you flip *PQRS* over in the left-right direction, you get the image in Figure 16-21.

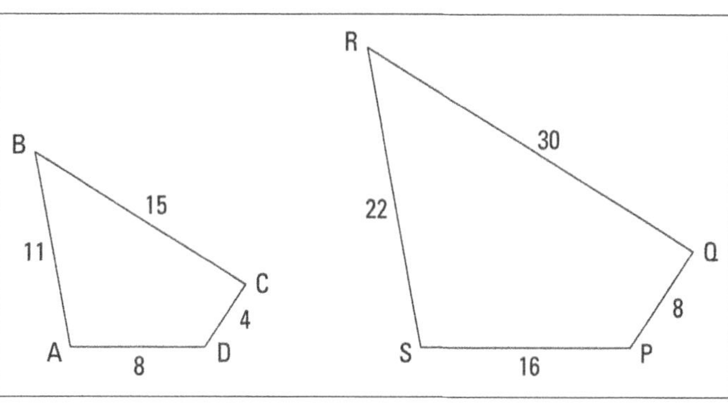

Figure 16-21:
Flipping *PQRS* over to make *SRQP* lines it up nicely with *ABCD* — pure poetry!

Now it's easier to see how the vertices pair up. *A* corresponds to *S*, *B* with *R* and so on, so you write the similarity like this: *ABCD ~ SRQP*.

Align similar polygons. If you get a problem with a diagram of similar polygons that aren't lined up, consider redrawing one of them so that they're both positioned in the same way. This may make the problem easier to solve.

Solving a similarity problem

Enough of this general stuff — let's see these ideas in action:

Given: *ROTFL ~ SUBAG*

          Perimeter of

          ROTFL is 52

Find: 1. The lengths of

          $\overline{AG}$ and $\overline{GS}$

          2. The perimeter

          of *SUBAG*

          3. The measures of

          $\angle S, \angle G,$ and $\angle A$

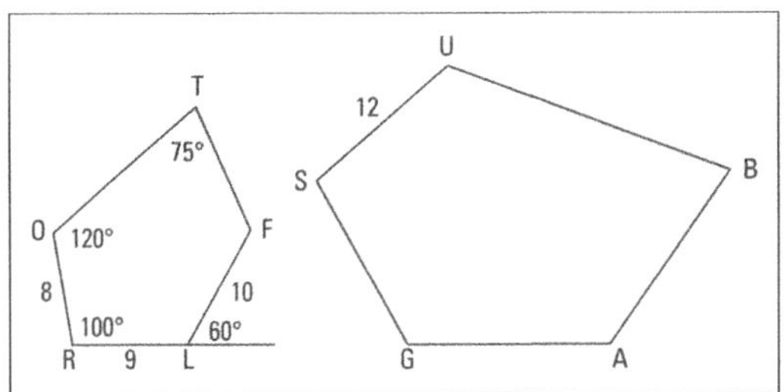

You can see that the *ROTFL* and *SUBAG* aren't positioned the same way just by looking at the figure (and noting that their first letters, *R* and *S*, aren't in the same place). So you need to make sure you pair up their vertices correctly, but that's a snap because the letters in *ROTFL ~ SUBAG* show you what corresponds to what. *R* corresponds to *S*, *O* corresponds to *U* and so on. (By the way, do you see what you'd have to do to line up *SUBAG* with *ROTFL*? *SUBAG* has sort of been tipped over to the right, so you'd have to rotate it counter-clockwise a bit and stand it up on base $\overline{GS}$. You may want to redraw *SUBAG* like that, which can really help you see how all the parts of the two pentagons correspond.)

Here's how to solve the problem:

1. **Find the lengths of $\overline{AG}$ and $\overline{GS}$.**

   The order of the vertices in $ROTFL \sim SUBAG$ tells you that $\overline{SU}$ corresponds to $\overline{RO}$ and that $\overline{AG}$ corresponds to $\overline{FL}$; thus, you can set up the following proportion to find missing length $\overline{AG}$:

   $$\frac{AG}{FL} = \frac{SU}{RO}$$

   $$\frac{AG}{10} = \frac{12}{8}$$

   $$\frac{AG}{10} = 1.5 \text{ (or you could have cross-multiplied)}$$

   $$AG = 15$$

   This method of setting up a proportion and solving for the unknown length is the standard way of solving this type of problem. It's often useful, and you should know how to do it.

   But another method can come in handy. Here's how to use it to find $\overline{GS}$: Divide the lengths of two known sides of the figures like this: $\frac{SU}{RO} = \frac{12}{8}$, which equals 1.5. That answer tells you that all the sides of *SUBAG* (and its perimeter) are 1.5 times as long as their counterparts in *ROTFL*. The order of the vertices in $ROTFL \sim SUBAG$ tells you that $\overline{GS}$ corresponds to $\overline{LR}$; thus, $\overline{GS}$ is 1.5 times as long as $\overline{LR}$:

   $$GS = 1.5 \cdot LR$$

   $$= 1.5 \cdot 9$$

   $$= 13.5$$

2. **Find the perimeter of *SUBAG*.**

   The method I just introduced tells you that

   $$\text{Perimeter}_{SUBAG} = 1.5 \cdot \text{Perimeter}_{ROTFL}$$

   $$= 1.5 \cdot 52$$

   $$= 78$$

3. **Find the measures of $\angle S$, $\angle G$ and $\angle A$.**

   S corresponds to R, G corresponds to L, and A corresponds to F, so

- Angle S is the same as $\angle R$, or $100°$.

- Angle G is the same as $\angle RLF$, which is $120°$ (the supplement of the $60°$ angle).

To get $\angle A$, you first have to find $\angle F$ with the sum-of-angles formula:

$$\text{Sum of angles}_{\text{Pentagon}ROTFL} = (n-2)180$$
$$= (5-2)180$$
$$= 540°$$

Because the other four angles of *ROTFL* (clockwise from *L*) add up to $120° + 100° + 120° + 75° = 415°$, $\angle F$, and therefore $\angle A$, must equal $540° - 415°$, or $125°$.

Proving Triangles Similar

In the section 'Proving Triangles are Congruent', I explain five ways to prove triangles congruent: SSS, SAS, ASA, AAS and RHS. Here, I show you the three ways to prove triangles *similar*: AA, SSS~, and SAS~.

Use the following methods to prove triangles similar:

- **AA:** If two angles of one triangle are congruent to two angles of another triangle, the triangles are similar.

- **SSS~:** If the ratios of the three pairs of corresponding sides of two triangles are equal, the triangles are similar.

- **SAS~:** If the ratios of two pairs of corresponding sides of two triangles are equal and the included angles are congruent, the triangles are similar.

Tackling an AA proof

The AA method is the most frequently used and is, therefore, the most important. Luckily, it's also the easiest of the three methods to use. Give it a whirl with the following proof:

Given: $\angle 1$ is supplementary

   to $\angle 2$

   $\overline{AY} \parallel \overline{LR}$

Prove: $\triangle CYA \sim \triangle LTR$

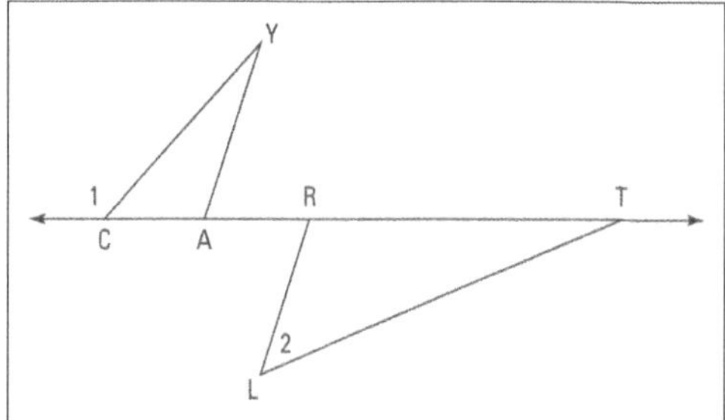

Here's a possible game plan (this hypothetical thought process assumes that you don't know that this is an AA proof from the title of this section): The first given is about angles, and the second given is about parallel lines, which will probably tell you something about congruent angles. Therefore, this proof is almost certainly an AA proof. So all you have to do is think about the givens and figure out which two pairs of angles you can prove congruent to use for AA. Easy peasy.

Take a look at how the proof plays out:

Statements	Reasons
1) $\angle 1$ is supplementary to $\angle 2$	1) Given.
2) $\angle 1$ is supplementary to $\angle YCA$	2) Two angles that form a straight angle (assumed from diagram) are supplementary.
3) $\angle YCA \cong \angle 2$	3) Supplements of the same angle are congruent.
4) $\overline{AY} \parallel \overline{LR}$	4) Given.
5) $\angle CAY \cong \angle LRT$	5) Alternate exterior angles are congruent (using parallel segments $\overline{AY}$ and $\overline{LR}$ and transversal $\overleftrightarrow{CT}$).
6) $\triangle CYA \cong \triangle LTR$	6) AA. (If two angles of one triangle are congruent to two angles of another triangle, then the triangles are similar; lines 3 and 5.)

Part V
The Part of Tens

In this part . . .

✔ Collect tricks to help you avoid making common algebra mistakes.

Chapter 17

Ten Ways to Avoid Algebra Pitfalls

In This Chapter

▶ Watching out for the pickle in the middle

▶ Distributing evenly and fairly for all concerned

▶ Making lemonade by putting a positive spin on subtraction

▶ Making it to first base with exponents

▶ Reducing fractions: Not a new diet

So much algebra is done in the world: Just about everyone who advances beyond primary school takes an algebra class, so the sheer number of people who use algebra means that a large number of errors are unavoidable. Forgetting some of the more obscure rules or confusing one rule with another is easy to do when you're in the heat of the battle with an algebra problem. But some errors occur because that error seems to be an easier way to do the problem. Not right, but easier — the path of least resistance. These errors usually occur when a rule isn't the same as your natural inclination. Most algebra rules seem to make sense, so they aren't hard to remember. Some, though, go against the grain.

The main errors in algebra occur while performing expanding-type operations: Distributing, squaring binomials, breaking up fractions or raising to powers. The other big error area is in dealing with negatives. Watch out for those negative vibes.

Keeping Track of the Middle Term

A squared binomial has three terms in the answer. The term that often gets left out is the middle term: The part you get when multiplying the two outer terms together and the two inner terms together and finding their sum.

The error occurs when just the first and last separate terms are squared, and the middle term is just forgotten.

Right	*Wrong*
$(a + b)^2 = a^2 + 2ab + b2$	$(a + b)^2 \neq a^2 + b^2$

Turn to Chapter 7 for more information on squaring binomials.

Distributing: One for You and One for Me

Distributing a number or a negative sign over two or more terms in parentheses can cause problems if you forget to distribute the outside value over every single term in the parentheses. The errors come in when you stop multiplying the terms in the parentheses before you get to the end.

Right	*Wrong*
$x - 2(y + z - w) = x - 2y - 2z + 2w$	$x - 2(y + z - w) \neq x - 2y + z - w$

You can find more on distributing in Chapter 7.

Breaking Up Fractions (Breaking Up Is Hard to Do)

Splitting a fraction into several smaller pieces is all right as long as each piece has a term from the numerator (top) and the entire denominator (bottom). You can't split up the denominator.

Right	*Wrong*
$\dfrac{x+y}{a+b} = \dfrac{x}{a+b} + \dfrac{y}{a+b}$	$\dfrac{x+y}{a+b} \neq \dfrac{x}{a} + \dfrac{y}{b}$

Go to Chapter 4 for more on dealing with fractions.

Renovating Radicals

If the expression under a radical has values multiplied together or divided, the radical can be split up into radicals that multiply or divide. You can't split up addition or subtraction, however, under a radical.

Right	Wrong
$\sqrt{a^2 + b^2} = \sqrt{a^2 + b^2}$	$\sqrt{a^2 + b^2} \neq \sqrt{a^2} + \sqrt{b^2}$

Note: The radical expression is unchanged because the sum has to be performed before applying the radical operation.

For more on radicals, turn to Chapter 6.

Order of Operations

The order of operations instructs you to raise the expression to a power before you add or subtract. A negative in front of a term acts the same as subtracting, so the subtracting has to be done last. If you want the negative raised to the power, too, then include it in parentheses with the rest of the value.

Right	Wrong
$-3^2 = -9$	$-3^2 \neq 9$
$(-3)^2 = 9$	

I fully discuss the order of operations in Chapter 5.

Fractional Exponents

A fractional exponent has the power on the top of the fraction and the root on the bottom.

When writing $\sqrt{x}$ as a term with a fractional exponent, $\sqrt{x} = x^{\frac{1}{2}}$. A fractional exponent indicates that a radical is involved in the expression. The 2 in the fractional exponent is on the bottom — the root always is the bottom number.

Right	Wrong
$\sqrt[5]{x^3} = x^{\frac{3}{5}}$	$\sqrt[5]{x^3} \neq x^{\frac{5}{3}}$

Check out Chapter 6 for more on fractional exponents.

Multiplying Bases Together

When you're multiplying numbers with exponents, and those numbers have the same base, you add the exponents and leave the base as it is. The bases never get multiplied together.

Right	Wrong
$2^3 \cdot 2^4 = 2^7$	$2^3 \cdot 2^4 \neq 4^7$

Turn to Chapter 6 for more on multiplying numbers with exponents and the same base.

A Power to a Power

To raise a value that has a power (exponent) to another power, multiply the exponents to raise the whole term to a new power. Don't raise the exponent itself to a power — it's the base that's being raised, not the exponent.

Right	Wrong
$(x^2)^4 = x^8$	$(x^2)^4 \neq x^{16}$

Chapter 6 is the place to go for more on powers.

Reducing for a Better Fit

When reducing fractions with a numerator that has more than one term separated by addition or subtraction, whatever you're reducing the fraction by has to divide every single term evenly in both the numerator and the denominator.

Right	*Wrong*
$\dfrac{(4+6x)}{4} = \dfrac{(2+3x)}{2}$	$\dfrac{(4+6x)}{4} \neq \dfrac{(2+6x)}{2}$

Go to Chapter 4 if you want more information on fractions.

Negative Exponents

When changing fractions to equivalent expressions with negative exponents, give every single factor in the denominator a negative exponent.

Right	*Wrong*
$\dfrac{1}{2ab^2} = 2^{-1}a^{-1}b^{-2}$	$\dfrac{1}{2ab^2} \neq 2a^{-1}b^{-2}$

You can find more on negative exponents in Chapter 6.

Publisher's Acknowledgements

We're proud of this book; please send us your comments through our online registration form located at `dummies.custhelp.com`.

Some of the people who helped bring this book to market include the following:

Acquisitions, Editorial and Media Development

Project Editor: Charlotte Duff

Acquisitions Editor: Kristen Hammond

Editorial Manager: Alice Berry

Production

Graphics: diacriTech

Technical Reviewer: Ingrid Kemp

Proofreader: Jenny Scepanovic

Indexer: Don Jordan, Antipodes Indexing

Every effort has been made to trace the ownership of copyright material. Information that enables the publisher to rectify any error or omission in subsequent editions is welcome. In such cases, please contact the Legal Services section of John Wiley & Sons Australia, Ltd.

Also available . . .

Index

• A •

AA proof of triangle similarity, 362–363
AAS (angle-angle-side) method of proving triangle congruency, 356
acute angles, 345
addends, 28
addition
  of decimal numbers, 80–82
  of different-signed numbers, 54
  of factors, 151–152
  of mixed numbers with different denominators, 75
  with powers, 144–145
  same-signed numbers, 52–54
  of variables, 143–144
  of whole numbers, 27–30
addition of fractions
  with different denominators, 68–75
  easy way, 69–71
  with same denominator, 67–68
  traditional way, 72–74
  using quick trick, 71–72
additive inverse numbers, 49
adjacent angles, 346
Aha algebra, 112
algebra
  applications of, 223–224
  vocabulary for, 11–12
angles
  definition, 341
  dividing, 348–349
  pairs of, 346–347
  in triangles, 333–334
  types of, 345
answers. *See also* solutions
  checking, 88–89
applications of algebra, 223–224
areas
  calculation methods, 320–324
  using formulae, 324–326

arguments as fun, 15
ASA (angle-side-angle) method of proving triangle congruency, 355
associative property of rules, 118–120
Australian Curriculum, 8
axes on Cartesian plane, 268–269, 270, 271

• B •

balance scale in counterfeit coin problem, 217
balancing equations with binary operations, 215–216
bases
  definition, 120
  multiplying, 370
binary operations, balancing equations with, 215–216
binomial base, exponents of, 123
binomials
  distribution of, 159–160
  factoring with, 208–209
  perfectly squared, 162–163
bisection
  of angles, 348
  of segments, 347
borrowing for subtraction, 33–36
bouncing ball problem, 136–138
boxes, volumes of, 327–328
braces, using, 99
brackets, using, 99

• C •

carrying over in addition, 28–30
Cartesian plane, 268–269
CAS calculators for factoring and expanding, 210
cat stalking mouse problem, 134–135
checking answers, 88–89, 105–107
checking solutions, 221–223

checks on operations, 223
circle
  area of, 324
  equation using graphs, 300–302
  graph of, 275–276
  perimeter of, 319–320
circumference, 319
coefficients, 141
collinear points, 342
column addition, 28
columns in subtraction, 32
combinations of operations, 147–149
common denominator for fractions,
  finding, 237
common factors, simplifying expressions
  by removing, 177–179
commutation of multiplication, 39, 153
commutative properties of rules, 117–118
complex fractions, 128
composite numbers
  definition, 114
  factorisation of, 169
composite shapes, areas of, 326
conceptual nature of numbers, 7
cone, volume of, 329
congruent triangles, proving, 350–357
connections between ideas, 15–16
constant, 11–12
constant term, eliminating in linear
  equations, 230–231
coordinates, 270–271
coplanar lines, 343, 344
coplanar points, 342
cosine (cos), 330
counterfeit coin problem, 217
counting numbers, 112
cubic equations, general form of, 214
cylinder, volume of, 328

decimal numbers
  addition of, 80–82
  completing division of, 87
  conversion to percentages, 90
  division of, 84–88

multiplication of, 83–84
  subtraction of, 82–83
definitions
  in geometry, 339–341
  precision in, 19
degrees of angle, 334–335
denominators, multiplying straight across,
  62–63
difference, 31
difference of two perfect squares,
  factoring, 207–208
different-signed numbers, adding, 54
distance between points on coordinate
  plane, 299–300
distributing
  problems in, 368
  to solve linear equations, 233–235
distribution
  of factors, 150–151
  of multiplication, 153
  of negative signs, 153
  of positive signs, 152
  of variables, 154–156
dividend
  definition, 43, 84
  trailing zeros in, 85–86
division
  before distributing, 235–237
  of decimal numbers, 84–88
  of fractions, 62–66
  of mixed numbers, 66
  of signed numbers, 56–58
  use in solving linear equations,
    225–227
  of variables, 146–147
  of whole numbers, 43–47
division table, 44
divisor, 43, 84
'Does this make sense?' question, 19–20
'dot' symbol for multiplication, 37

eighth index law, 129–131
encryption and prime numbers, 167
equal shares, 149–150

equations
  algebraic, 195
  balancing with binary operations, 215–216
  definition, 11
  setting-up for solving, 214
  solving with reciprocals, 219–220
  squaring terms in, 217
equilateral triangles, 328, 350
expanding, 117
expansion of expressions, 149–152
exponential expressions, 133–138
exponentiation, 120
exponents
  of binomial base, 123
  definition, 12, 120–121
  negative, 128, 371
expressions. *See also* order of operations;
    order of precedence
  algebraic, 195
  definition, 11
  expansion of, 149–152
  removing common factors for
      simplification, 177–179
expressions with fractions as exponents,
    changing radicals to, 158
extra variable term, removing to solve
    linear equations, 231–233

• F •

factoring
  with binomials, 208–209
  changing to a division problem, 185–186
  completion of, 209–210
  composite numbers, 169
  definition, 117
  difference of two perfect squares,
      207–208
  as division, 180–181
  grouping terms for, 186–190
  outlining method of, 181–182
  in the real world, 182
  repeated-division method, 181–182
  for solutions to quadratics, 249–252
  terms and rules for, 180
  to solve trinomial equations, 253–257

factoring out
  combinations of numbers and variables,
      183–185
  numbers, 180–182
  variables, 182–186
factors
  addition of, 151–152
  definition, 36
  distribution of, 150–151
fifth index law, 126–127
first index law, 121–123
flash cards for multiplication, 39–40
FOIL acronym, 195
FOILing
  applying to a special product, 198–199
  basics of, 194–195
  to multiply two binomials together,
      196–198
formative assessment, 22–23
formulae, solving for variables in,
    241–242
fourth index law, 126–127
fraction line, 99
fractional equations, changing into
    proportions, 239–240
fractional exponents
  changing radical form to, 130
  multiplication of, 157–158
  pitfalls with, 369–370
fractions
  addition of, 67–75
  breaking up, 368
  changing negative exponents to, 156–157
  conversion to percentages, 91–92
  division of, 65–66
  reduced, 62–63
  reducing, 371
  reducing for prime factorisation, 174–177
  removing from equations, 237–239
  subtraction of, 75–80
fractions with different denominators
  addition of, 68–75
  subtraction of, 76–80
fractions with same denominator
  addition of, 67–68
  subtraction of, 76

functions
definition, 308–309
families of, 310–311
practical side of, 309
and variables, 310, 311

• G •

geometry, definitions in, 339–341
geometry proofs
definition, 337–338
using knowledge of, 339
gradient
combining with intercept, 289–290
formula for calculating, 287–289
of straight-line graph, 285–287
gradient-intercept form of line equation,
289–291
graphs
labelling points on, 272
points on, 269–270
value of, 273–274
greatest common factor (GCF)
in quadratic expressions, 193–194
in solving quadratic equations, 250–252
use of, 177
using with unFOILing to factor
quadratics, 204–206
grouping, using with unFOILing to factor
quadratics, 206–207
grouping rules, combined with order of
operations, 103–105
grouping symbols
in algebra, 99–105
as instructions, 115–116

• H •

Heron's formula, 323–324
homework
helping children with, 22–23
teachers' views on, 24
homework assignments, problems with,
24–25
horizontal lines, 343

'How do you know?' question, 17–18, 23
hypotenuse, 314–315

• I •

ideas, connections between, 15–16
imaginary numbers, example of, 248
improper fractions, 64
index laws
eighth, 129–131
first, 121–123
fourth, fifth and sixth, 126–127
second, 124–125
seventh, 128
third, 126
indexes (indices), 120–121
infinite geometric sequence, 136–137
integers, 113
intercept, combining with gradient,
289–290
intercepts on graph axes, 284–285
intersecting lines
definition, 343, 344
graphing, 293–294
intersecting planes, 344
intersections
eliminating to find, 296–298
subsituting to find, 294–296
inverse operations, 44
irrational numbers, 114
'Is it good enough?' question, 18–19
isosceles triangles, 350

• L •

language, precision in use of, 19
larger numbers, multiplication of, 41–43
least common multiple (LCM), 73
length
calculating perimeters, 316–320
of sides of triangles, 330–333
units of, 314
light year, 132
line graphs, plotting, 279–281
line segment, 340

linear equations
  general form of, 214
  graphing, 274–275, 281–284
  order of operations for solving, 224–225
  simplifying, 233–237
  solving by distributing, 233–235
  with three terms, 230–233
  with two terms, 225–230
lines. *See also* rays; segments
  definition, 340
  terms for pairs of, 343–344
long division, 44–47

• *M* •

making sense of answers, 105–107
'making sense' question, 19–20
maths, playing with, 13–16
measurements, 313–320
metric system, 314
middle terms, keeping track of, 367–368
midpoint between two points on
    coordinate plane, location of, 299–300
minuend, 31
minus symbol (–), 141–142
minutes of angle, 334–335
mixed numbers
  division of, 66
  multiplication of, 64–65
mixed numbers with different
    denominators
  addition of, 75
  subtraction of, 79–80
mixed-operator expressions, order of
    operations in, 97
models, building, 14
multiplicand, 36
multiplication
  before distributing, 235–237
  of decimal numbers, 83–84
  distribution of, 153
  of fractional exponents, 157–158
  of fractions, 62–66
  of larger numbers, 41–43
  of mixed numbers, 64–65
  of signed numbers, 56–58

use in solving linear equations, 227–229
  of variables, 145–146
  of whole numbers, 36–43
multiplication 'dot' symbol, 37, 122
multiplication property of zero (MPZ)
  definition, 249
  in solving quadratic equations, 250–252
multiplication table, memorising, 37–41
multiplicative identity, 39
multiplier, 36

• *N* •

natural numbers, 112
negative exponents, 128
  changing to fractions, 156–157
  pitfalls with, 371
negative numbers
  compared with positive numbers, 51–52
  definition, 50–51
  indicating, 49
negative signs, distribution of, 153
9 times table, memorising trick, 41
non-collinear points, 342
non-coplanar lines, 343–344
non-coplanar points, 342
Null Factor Law (NFL)
  statement of, 249
  use in solving quadratic trinomials,
    253–257
numbers
  in algebra, 111–114
  conceptual nature of, 7
  factoring out, 180–182
  representing with letters, 140–142
numerators, multiplying straight across,
    62–63

• *O* •

oblique lines, 343, 344
obtuse angles, 345
one
  in fraction problems, 64
  not a prime number, 168

operations
  checks on, 223
  combinations of, 147–149
  definition, 11
order of operations, 93–95
  combined with grouping rules, 103–105
  expressions with addition and subtraction, 95–96
  expressions with exponents and roots, 98
  expressions with multiplication and division, 96–97
  pitfalls with, 369
  for solving linear equations, 24–225
order of precedence
  in expressions with exponents and parentheses, 100–101
  in expressions with nested parentheses, 102
  in expressions with parentheses, 100
  in expressions with parentheses raised to an exponent, 101–102
ordered pairs of numbers, 270–271

• P •

palindromes, 154
parabolas
  basic, 304–305
  definition, 303–304
  graphs of, 276–277
  putting vertex on an axis, 305
  sliding and multiplying, 305–308
parallel lines
  definition, 343, 344
  gradients of, 292–293
parallel planes, 344
parentheses, 99
parents, help by, 17–25
percentages
  conversion to decimal numbers, 89
  conversion to fractions, 90–91
perfectly squared binomials, 162–163
perimeter, calculating length of, 316
perpendicular lines
  definition, 343, 344
  gradients of, 292–293

physics, quadratic equations in, 257–258
π, 115–116, 319
planes
  definition, 341
  pair of, 344
plus symbol (+), 141–142
points
  definition, 340
  types of, 342
points on Cartesian plane, plotting, 272
points on graphs, 269–270
polygons
  perimeters of, 318
  similar, 357–358
polynomials, multiplication by another polynomial, 161
positive numbers
  compared with negative numbers, 51–52
  definition, 50
positive signs, distribution of, 152
power of zero, 126
powers
  function of, 120–121
  raised to a power, 126–127, 370
powers of a variable, factoring out, 182–183
precision of algebraic symbols, 141–142
prime factorisation
  of a number, 169
  reducing fractions for, 174–177
  rules of divisibility for, 172–174
  using a tree, 171–172
  using upside-down division, 170–171
prime numbers
  definition, 114, 168–169
  trinomials as, 203–204
principal square root of a number, 247
prisms, volumes of, 327–328
product of sum and difference of same two terms
  applying FOIL to, 198–199
  method, 163–165
profit calculations, quadratic equations in, 258–259
proofs, in geometry, 337–338
proofs. See geometry proofs

proportions, solving equations using, 237–239
pyramids, volumes of, 328–329
Pythagoras, life of, 316
Pythagorean formula, using, 314–316

• Q •

quadrants on Cartesian plane, 268–269, 271
quadratic, origin of word, 250
quadratic equations
  applications of, 257–259
  factoring for a solution, 249–252
  general form of, 214
  multiple answers to, 244–246
  solving when b = 0, 246–248
quadratic equations with three terms, solving, 252–257
quadratic expressions
  choosing factoring technique, 204–206
  definition, 191
  simplifying, 193–194
  standard form on, 192–193
quadratic formula, 259–263
quadratic trinomial equations, solving by factoring, 252–257
questions, asking by parents, 17–22
quotient, 43, 84

• R •

radical form, changing to fractional exponent, 130
radicals
  changing to expressions with fractions as exponents, 158
  definition, 99
  and roots, 129–131
  splitting, 369
ratio, in infinite geometric sequence, 136–137
rational numbers, 114–115
rays. *See also* lines; segments
  definition, 340
real numbers, 112

reality checks on solutions, 221–222
reciprocals
  in division of fractions, 65–66
  of numbers, 128
  solving equations with, 219–220
  use in solving linear equations, 229–230
rectangle
  area of, 321
  perimeter of, 318
reduced fractions, 62–63
reflex angles, 345
regular pentagon, 328
Rhind Mathematical Papyrus, 112
right angles, 345
right-angled triangle, 314
rise over run, 287
root power, 129
roots
  and radicals, 129–131
  taking in solving equations, 218
rounding off answers, 87–88
rules
  associative property of, 118–120
  commutative property of, 117–118
  for operating with exponents, 120–121
  for radical expressions, 130
rules of divisibility for prime factorisation, 172–174

• S •

same-signed numbers, addition of, 52–54
SAS (side-angle-side) method of proving triangle congruency, 353–355
scalene triangles, 349
scientific notation of numbers, 131–133
second index law, 124–125
segments. *See also* lines; rays
  bisection and trisection, 347–348
  definition, 340
seventh index law, 128
shapes
  structure of, 20–22
  using knowledge of, 338–339
short multiplication table, 38–40

signed numbers
  multiplying and dividing, 56–58
  subtracting, 54–56
  and zero, 58–59
similar figures, lining up to prove
    similarity, 358–360
similar polygons, 357–358
similarity of figures, 357–362
similarity problems, solving, 360–362
similarity of triangles, proving, 362–363
simplifying, 116
simultaneous equations, applications of,
    298–299
sine (sin), 330
sixth index law, 126–127
solutions. *See also* answers
  checking, 221–223
  reality checks on, 221–222
solving, 117
sphere, volume of, 329–330
square
  area of, 321
  perimeter of, 317–318
squaring terms in equations, 217
SSS (side-side-side) method of proving
    triangle congruency, 351–353
stacks in subtraction, 32
statement, definition, 195
story problems, 117
straight angles, 345
structure in maths, 20–22
subtraction
  of decimal numbers, 82–83
  of fractions, 75–80
  of fractions with different denominators,
    76–80
  of fractions with same denominator, 76
  with powers, 144–145
  of signed numbers, 54–56
  of variables, 143–144
  of whole numbers, 31–36
subtraction of fractions
  easy way, 77–78
  traditional way, 78–79
  using quick trick, 78
subtrahend, 31

summing. *See* addition
supplementary angles, 347
surd form, 262
symbols
  in algebra, 115
  experimenting with, 13–14
  meaning of, 141–142
  for relationships, 116

• T •

taking away. *See* subtraction
tangent (tan), 330
terms, 11, 149
third index law, 126
3-D (three-dimensional) space, 341
times. *See* multiplication
trailing zeros, in dividend, 85–86
tree, using to find prime factorisations,
    171–172
triangles
  areas of, 322–324
  equilateral, 350
  isosceles, 350
  lengths of sides, 330–331
  perimeters of, 317
  proving congruency of, 350–357
  proving similarity of, 362–363
  scalene, 349
trigonometric ratios, 330–335
trinomial equations, factoring, 253–257
trinomials
  distribution of, 160–161
  as primes, 203–204
trisection, of angles and segments, 348
two perfect squares, factoring the
    difference of, 207–208

• U •

undoing operation with its opposite, 218
unFOILing to factor quadratics
  process, 199–204
  using with GCF, 204–206
  using with grouping, 206–207

units, attention to, 19
upside-down division, 170–171

 • _V_ •

variables
  distribution of, 154–156
  division of, 146–147
  factoring out, 182–186
  multiplication of, 145–146
variables in algebra, 11, 12, 140
variables in formulae, solving for, 241–242
vertex of parabola
  definition, 303–304
  moving, 305–308
  putting on an axis, 305
vertical line test, 308–309
vertical lines, 343
vinculum, 99
volumes, calculating, 327–330

 • _W_ •

'What if?' questions, 18
'What's going on here?' question, 20–22
whole numbers, 113
'Why?' questions, 17–18

 • _X_ •

x-y coordinate system, 268–269

 • _Z_ •

zero
  as answer to sums, 56
  function of, 52
  power of, 126
  property of multiplication, 39, 249–250
  and signed numbers, 58–59

Printed and bound by CPI Group (UK) Ltd, Croydon, CR0 4YY

07/07/2026

14916213-0001